THE MOBILE ERA OF INTENT

Six Pillars for Intent-Driven AI

Van E. Eiseman

EISEMAN PUBLISHING

ISBN 979-8-9930233-0-4

First Trade paperback edition 2026

Editing by Peter Letzelter
Proofreading by Jesse Ackles
Cover design by Van E. Eiseman
Text design and composition by Gena Eiseman

Dedicated to

Gena, Maya, and Lance

Table of Contents

Table of Contents

Table of Contents

Introduction -

The Choice Point

The Choice Point

In nearly three decades of working across digital, mobile, and AI systems, I've learned to pay attention, to spot patterns when the pace of advancement seems to surpass the clarity of its underlying purpose. Over the past few years, that gap between velocity and assessment has widened in ways I find increasingly hard to ignore.

New AI capabilities arrive almost daily, yet the reasoning behind key design choices—what data is collected, why systems require certain access, how predictions influence behavior—often remains hidden. Some systems genuinely support human capability. Others introduce trade-offs without acknowledging them. A few extend into areas of autonomy or influence that most people never explicitly agreed to.

These shifts aren't inherently good or bad. They are directional signals. And when direction is unclear, the outcomes that eventually emerge are shaped less by intentional design and more by momentum.

Early in developing what I named the **Mobile Era of Intent**—the threshold moment when technology can finally understand human intent rather than forcing human adaptation to machine logic—I found myself returning to the same directional questions: Why was this system designed this way? What intent shaped its architecture? What assumptions guided its behavior? And the question at the center of all the others: Should it be built or deployed in this form at all?

The more I explored these questions, the more I realized they weren't appearing in most AI discussions. Instead, what was evident across topic after topic was several key insights. Strategic leaders were under pressure to keep pace with competitors. Product teams were navigating complex new capabilities without clear frameworks for assessing impact. Policymakers were addressing harms only after they had emerged, rather than guiding choices before they solidified into standard practice. And individual users were asked to trust systems designed with incentives they had little ability to understand or influence.

The absence of these questions in public AI discourse wasn't trivial. It had real-world consequences. Many AI systems are marketed as helpful and friendly, but they optimize for engagement metrics, attention capture, or data extraction. For most people, these pressures appear as subtle psychological pulls. But in extreme cases—such as the tragic pattern highlighted by the case of Sewell Setzer III, a teenager who died by suicide after intensive interaction with an AI companion chatbot—misaligned AI systems can magnify vulnerability into devastating harm. These are not speculative risks; they are the predictable results of systems designed without clear evaluative criteria for how technology should operate.

I wanted to address something immediate: the choices being made right now, when those choices still determine direction. The rapid pace of AI adoption means today's design choices are quickly becoming tomorrow's technological infrastructure. There is a narrow window during which the technological direction can still be shaped. I believe we are in that window.

The central issue isn't the speed of AI development. The more pressing issue is the absence of shared evaluation frameworks. Without them, leaders can default to what's easy to measure. Teams might default to what's easy to ship. Policymakers could default to what's easy to regulate. And individuals often default to what's easy to accept, simply because there are few alternatives available.

The **Six Pillars of Intent** framework emerged from asking what kind of structure would enable these decisions to be made more consciously. They are not technical requirements or ideological questions. They are evaluation criteria, a way to assess clearly what is otherwise difficult to articulate: whether a system enhances human capability or nudges it aside; whether it strengthens or erodes the trust it requires; whether it respects human intent or subtly redirects it for its own purposes; whether it provides equitable access or concentrates advantage; whether it uses resources responsibly or shifts its costs to communities and the environment.

These pillars give strategic leaders the ability to evaluate investments and deployments based on human-centered alignment rather than on speed or competitive pressures. They give product teams language to distinguish enhancement from extraction when the difference isn't clear. They give policymakers a structure for examining incentives before harm reaches scale. And they give individual users—those with the least power yet the most at stake—a way to understand the forces shaping their relationship with the technology they engage with more and more every day.

What began as an attempt to describe a technological shift grew into an effort to articulate a framework for conscious choice. This isn't a prediction about where AI is headed or an argument for optimism or pessimism. It's a call to consciously evaluate the choices that determine which technological futures remain available. The Mobile Era of Intent describes a convergence already underway—the point where technology becomes capable of understanding and acting on human intent with increasing sophistication. What it does with that capability, and what we choose to do with it, depends on decisions being made today.

If you are someone responsible for shaping technology within an organization, I hope this framework helps you make decisions with greater confidence and more clarity. If you design or build these systems, I hope the pillars give you a vocabulary for advocating solutions that enhance human capability rather than replace it. If you actively shape or govern policy, I hope this approach helps illuminate where guardrails are most needed and most effective. And if you are an individual user simply trying to understand the world unfolding around you, I hope these chapters help you see the landscape with greater clarity.

The Six Pillars of Intent do not guarantee outcomes. They provide a way to recognize the choices that matter, the ones that determine whether AI strengthens human flourishing or undermines it. The framework that follows is one approach to evaluating those choices before they become embedded in infrastructure, business models, and societal patterns that are far harder to unwind later.

If this book provides a more transparent lens for examining the AI systems shaping your work, your decisions, and your digital environment, then it has fulfilled its *intent.*

Chapter 1 -

Agency or Inevitability

Agency

I was 17 when I learned that institutions don't much care about a person's dreams.

For four years, I'd known exactly what I wanted: to become an illustrator. Comics, perhaps. Brand advertising, maybe. Animation, for sure. Though the specifics didn't matter as much as the certainty of the plan: build a strong portfolio, graduate from high school, and attend college to find my path from there. My guidance counselor supported it. My art teacher championed it. By junior year, I had seven pieces I was proud of—original drawings, paintings, and logo designs that represented countless hours of work.

We packaged them carefully and sent them to a prestigious art school, along with a scholarship application coordinated by our school's admin office. This was my shot, my pathway to controlling my own future.

Six weeks later, I still hadn't heard back. When my counselor called the university, the admissions office delivered news that floored me: They never received the portfolio.

But the delivery had been confirmed. When pressed, the art school said they weren't sure where the work was and, regardless, they couldn't return it. They explained it wasn't their policy to return submitted artwork.

My best work ... gone. Four years of effort were lost in an institutional void. I blamed no one but myself. I should have made copies. I should have called the admissions office myself to learn their process before sending. But I was seventeen and trusted that following the guidance I'd been given would be enough.

I was devastated. The path I'd carefully constructed lay in ruins, and the people I counted on offered nothing more than policy statements and blame-shifting. I could have accepted defeat, settled for a mediocre art program at a local university along with a part-time job or two, and lived safely within the limitations I now faced.

I chose a more challenging path.

That summer, while waiting for my friend at a military recruiting station, a Navy recruiter asked about my plans after high school. When I mentioned art school and concerns about paying for it, he pitched an alternate route: four years of military service, followed by college "fully" funded by the GI Bill. It meant delaying my dreams, leaving my loved ones, and taking a much more difficult road to the same destination.

So, I enlisted in the U.S. Navy through the Delayed Entry Program. It was my conscious choice. My way of taking back control from a system that hadn't fully supported me. I chose agency—the human capacity to choose and direct outcomes—even when circumstances suggested otherwise. When the situation felt inevitable, agency offered a path forward.

Thirty-seven years later, I see similar patterns in how organizations choose to deploy technology. Not about art school or military service, obviously, but about whether decision makers will consciously shape how artificial intelligence serves human flourishing, or passively adopt deployment strategies focused solely on corporate benefit.

Major corporations are deploying AI systems without assessing the human impact. Amazon's AI systems have drawn scrutiny for uses beyond robots that optimize operations, including reports that algorithmic tools are being deployed to identify and isolate workers perceived as potential union organizers. Meta's platforms have faced ongoing criticism as engagement algorithms continue to amplify AI-generated spam and misinformation, following an attention-capture design that prioritizes emotionally charged content over accuracy.

What I see in these examples is institutional indifference to how this affects people: deploy first, wait for it to break, fix later. These patterns aren't unavoidable. They're the result of deliberate decisions that prioritize efficiency over evaluation, automation over impact assessment, and speed to market over societal benefit.

I believe this is a crucial moment where these decisions matter more than they have in decades. The systems being deployed today have the potential to shape how billions of people work, communicate, learn, and connect for generations to come. People can participate in those design choices, or they can passively accept whatever emerges from competitive pressure and market forces.

When my portfolio disappeared into a bureaucratic black hole, I learned that circumstances don't determine outcomes—human agency does. I see the same principle at work in the AI shift happening now. Yet most people experience the current AI trajectory as inevitable.

Inevitability

Inevitability can feel real when there's no clear, conscious choice available. Media headlines amplify this perception with dramatic predictions about AI transformation, often designed more for engagement than accuracy. Corporate earnings calls frequently justify rushing AI deployment as a competitive necessity. Venture capital flows to AI startups regardless of proven utility. Regulatory discussions often lag months behind technological capabilities, creating a vacuum that companies interpret as permission to proceed.

This creates a false sense of inevitability. The AI trajectory we're witnessing isn't the direct result of natural tech progress. It's primarily the outcome of specific choices made by particular actors, who prioritize speed over deliberation, market capture over human impact assessment, and automation over augmentation.

But even false inevitability is powerful. When technology deployment patterns feel unstoppable, people tend to focus on adaptation rather than participation in design decisions. Recent research shows that while 60% of workers expect AI to change their tasks significantly by 2030, only 39% feel involved in shaping workplace AI solutions. 62% of consumers express unease about AI handling their data, yet 37% continue sharing information simply because they feel there's no alternative. These patterns reinforce people's feelings that AI development is happening *to* them, not *with* them.

Equally, the speed of deployment creates its own momentum. Companies rush to integrate AI capabilities before competitors, often without clear use cases or established success metrics. This creates a feedback loop where AI adoption becomes self-perpetuating. Not because it solves defined problems, but because everyone else is doing it.

I've observed this pattern before. During the dot-com boom, companies added ".com" to their names without reevaluating their business models. During the mobile revolution, organizations built apps that replicated their websites, failing to leverage what mobile could actually do. Today, I observe a similar trend with AI integration: deployment for its own sake, driven by competitive pressure rather than purposeful implementation.

The difference this time is scale and consequence. Unlike other waves of technology that mainly affected specific industries or demographics, AI deployment is beginning to touch every aspect of how people work, communicate, and access information. The stakes of getting it wrong, or right, are exponentially higher.

This is why the choice between agency and inevitability matters so much right now. The patterns established in the next few years will likely persist for decades, shaping how billions of people interact with more advanced systems. Organizations and individuals can choose conscious implementation guided by human flourishing, or accept whatever emerges from competitive pressure and automation. The choice should be obvious, yet many still default to accepting what feels inevitable.

But this choice is complicated by a deeper dynamic: the role of trust itself.

The Trust Paradox

What if I were to tell you that AI is an alien technology? Imagine discovering ancient scrolls that, when deciphered, explained sophisticated patterns for understanding any language, making requests, and even new forms of communication between humans and the cosmos.

Most rational people would hesitate before trusting such powerful systems we can't understand, regardless of their potential. Yet this mirrors exactly what historian Yuval Noah Harari identifies as "*the trust paradox*" at the heart of AI development: technology leaders justify rapid deployment by citing competitive pressure—if they don't build it, someone else will. They don't trust their human competitors to develop AI responsibly. Yet these same leaders ask billions of people to trust the AI systems they're creating to remain aligned with human interests.

Harari points to a basic contradiction in this approach. As he observes, "We have thousands of years of experience with human beings. We have a broad understanding of human psychology and biology, of the human craving for power, and of the forces that keep the pursuit of power in check." Humanity has spent thousands of years building institutions, social norms, and governance structures that enable cooperation despite conflicting interests.

With AI, however, the same developers who won't trust their competitors are asking society to trust artificial systems that can learn, adapt, and pursue goals independently. "We already know that even primitive AI systems can lie, manipulate, and adopt goals and strategies not foreseen by their human developers," Harari notes. Yet people are being asked to extend trust to systems whose decision-making processes can't be fully explained or verified.

This paradox becomes even sharper when considered in reverse: If beings with vastly superior intelligence were to arrive, offering to solve humanity's challenges but revealing nothing about their motivations or values, most would recognize the danger, regardless of their own capabilities. Yet AI systems that will soon surpass human abilities in most domains are increasingly being entrusted with decisions about health, opportunities, relationships, and futures. At the same time, decision makers understand very little about how these systems work or what principles guide their choices.

Harari concludes that building more trust between humans must precede the development of truly superintelligent AI systems. His reasoning is straightforward: "Together, humans can control AI. But if we fight one another, AI will control humanity."

His insight highlights the central tension of our technological era. However, it also raises a deeper question about how trust operates when adopting new technology. The current AI trajectory assumes trust is binary—either people trust these systems, or they don't.

But my observation from contributing to digital progress over the past 30 years suggests that trust operates more like a complex investment than an on/off switch. People invest trust in technologies based on perceived returns. Those investments can either compound through demonstrated value or be squandered through misaligned incentives. But it's not often so cut and dry.

Sometimes distrust serves as productive feedback rather than an obstacle to overcome. When workers express concerns about workplace AI, when consumers hesitate to share data, or when policymakers call for regulation, that skepticism often signals a genuine misalignment between what tech can do and what people need it to do. Distrust can inform better design choices when companies treat it as feedback rather than user resistance to be mitigated.

Understanding these trust dynamics becomes crucial for addressing Harari's paradox. The same competitive pressure that prevents cooperation between AI developers also shapes how people decide whether to trust the systems those developers create. Without clear ways to evaluate whether AI systems serve human intent or exploit it, trust becomes weaponized—exactly what Harari warns against.

The Currency of Progress

Over the past four decades, I've witnessed people share personal information with early email systems not because they fully trusted the security, but mostly because email felt inevitably the best way to stay connected. I've seen users sign up for social media platforms they had reservations about because that's where most of their families and friends were gathering. I've observed businesses adopting cloud services despite many security concerns, as competitors gain advantages through digital transformation.

These observations align with Harari's insight that we must first establish trust among humans. However, there's additional complexity that helps explain why building this human trust proves so challenging in practice.

I believe the difficulty lies in the complex relationship between human trust and innovation. Companies that deliver genuine value often earn deeper trust and enthusiastic adoption. But many also succeed simply by becoming the default option, benefiting from user inertia rather than earning confidence. Similarly, companies that exploit user trust often do not face immediate consequences. Users may continue to use services they distrust because of network effects, convenience, or a lack of viable alternatives.

These observations reveal that people make investment calculations when considering new technologies, weighing questions like: *What am I giving up? What can I expect to receive in return? Is this trade-off worth the risk?*

This led me to think differently about the relationship between trust and technological progress: Trust in information functions as currency—it can pay for genuine progress or accumulate as technological debt.

These trust investments, what I call "*trust as currency*," can take multiple forms simultaneously. Sometimes conscious and calculated—weighing specific benefits against known risks. Sometimes passive and resigned—accepting technologies because alternatives feel inadequate or unavailable. Often, they're a mixture of both, with enthusiasm for certain features coexisting with unease about others, leading to begrudging trade-offs or resignation to systems that serve companies more than users.

Understanding how this currency gets invested matters for worthwhile innovation. When people choose to invest trust in a technology company, they're betting that their investment will yield returns—enhanced experiences, protected privacy, mutual benefit, or some other form of value exchange.

Companies can honor these investments or extract trust without providing comparable returns. When they extract more than they give back, they're essentially borrowing against their future—creating technological debt that eventually demands payment through user backlash, regulatory intervention, or lost operational freedom.

How companies handle these trust investments determines whether they build sustainable innovation or burn through user goodwill. Two companies illustrate this dynamic clearly.

Netflix provides the clearest example of that trust investment, building sustainable innovation. When the company announced its shift from DVD mail rental to streaming in 2007, Netflix asked people to believe that reliable internet streaming was possible, that content libraries would expand rather than shrink, and that smart recommendations could meaningfully improve content discovery. Subscribers faced a fundamental choice: Could they trust that this new technology would deliver better value than getting DVDs in the mail?

Netflix honored the trust by solving enormous technical challenges—streaming to millions without buffering and offering personalized suggestions that improved discovery—in service of the user experience, rather than exploiting user behavior for additional revenue streams. They reinvested streaming revenue in improving content quality and platform reliability. Their recommendation system improved content discovery rather than manipulating viewing habits to maximize engagement metrics. They maintained transparent pricing and didn't sell viewing data to third parties for advertising revenue.

Netflix's approach paid massive dividends. Subscribers gained unprecedented convenience, a wide variety of content, and personalized discovery. Netflix demonstrated that honoring trust creates a sustainable competitive advantage, as its subscriber base grew from 7 million in 2007 to over 300 million globally by 2025.

Facebook embodies the opposite dynamic: the systematic squandering of trust for corporate gain. When millions of people entrusted Facebook with their personal information, relationship data, and social connections, they did so on the promise of enhanced human connection and a sense of community.

However, when investors pressured Facebook to monetize, user trust was exploited rather than honored. The company collected far more personal data than necessary for social connection, used psychological profiling to influence behavior rather than simply connecting people, and optimized user feeds for engagement metrics that often amplified divisive content. The Cambridge Analytica scandal epitomized this exploitation—Facebook's systems allowed political consultants to harvest personal data from millions of users without consent, repurposing social connection data for political manipulation.

Despite well-documented privacy violations and psychological manipulation, hundreds of millions of people continue using the platform—not because they trust Facebook's intentions, but because they feel trapped by network effects and limited alternatives. Users endure Facebook's practices because, as many express it, "it's the only place everyone else is."

This unconscious acceptance complicates the trust as currency exchange. Netflix benefits from both earned loyalty and market dominance. Facebook profits from both active users and those who feel they lack acceptable alternatives.

In both cases, the trust investment encompasses not only conscious choice about value exchange but also practical surrender to market realities and social pressures.

The AI deployment patterns I observe today echo these same complex dynamics. Some companies are developing AI systems that augment human capabilities and preserve user agency, adopting a model that respects trust while also benefiting from market positioning. Others are deploying AI to optimize behavioral manipulation and data extraction, following a model of exploiting trust while relying on user resignation and limited alternatives.

However, unconscious acceptance risks more consequential choices than those in previous technology eras. When AI systems are adopted into society, whether enthusiastically or reluctantly, they gain unprecedented access to how people behave, make decisions, and connect socially. Unlike streaming services and social media platforms, where users can observe what they consume or share, AI systems often operate behind the scenes—processing everyday data and influencing choices with limited user visibility.

Understanding how trust functions as currency becomes essential as AI systems develop more sophisticated capabilities to interpret human intent and influence human behavior. The trust people invest in these technologies today—conscious and unconscious, willing and resigned, and everything in between—shapes the current AI trajectory. But this raises a fundamental question: what problems does AI solve with these trust investments?

Is AI Solving the Right Problem?

When philosopher Michael Sandel asks, "To what problem is AI the solution?" he challenges us to consider whether we're applying this transformative technology to enhance human capability and community, or simply to optimize corporate efficiency at society's expense. His question highlights a fundamental misalignment in current AI deployment: many companies are applying AI to solve efficiency problems rather than the human-centered problems that would honor the trust invested in technological progress.

I experienced this contrast firsthand when building a chatbot for my website. Creating a fully functional AI system with CRM integration would have required weeks of development work just a few years ago. Instead, I built a working lead-generation system over a weekend. This represents genuine augmentation—AI solving a real problem I actually had, enabling me to learn and build it myself rather than offloading the experience.

But purposeful applications like this make up a small fraction of current AI deployments. Consider what's actually scaling across corporate America in 2025. Major consulting firms sell "automation agent workforces" to replace customer service representatives and claims processors. Companies deploy chatbots to handle complaints without addressing why customers are frustrated in the first place. Healthcare systems use AI to process insurance claims more efficiently, even though those claims would otherwise take weeks to resolve.

This rush to automate business processes represents choices that avoid the hard work real problem solving requires. Instead of using AI to expand who gets coverage, companies automate claim processing to make administrative tasks more efficient. Instead of examining why customer service interactions frustrate people, chatbots eliminate human touchpoints without necessarily improving outcomes.

The contrast becomes clearer when examining purposeful deployment. Cities such as Oslo, Norway, use AI for urban planning—analyzing traffic patterns, optimizing infrastructure, and addressing genuine citizen needs to improve transportation and services. This represents problem solving that honors technology's potential rather than extracting efficiency from broken systems.

Workforce responses put these deployment choices in perspective. Hollywood writers went on strike for AI protections because their creative work was being used to train systems that could reduce demand for human creators—without consent or compensation. Healthcare workers pushed back against AI diagnostic tools because they understood the stakes of replacing human judgment in critical decisions. European unions demanded workplace AI regulation because deploying automation without worker input constitutes a fundamental breach of trust.

These responses aren't resistance to technology—they're a recognition that most current AI deployments don't solve the problems people actually experience. Worker skepticism, regulatory scrutiny, and consumer wariness serve as productive signals when companies are willing to listen. Most workers aren't asking for job elimination. Most customers aren't requesting more automated service interactions. Most citizens aren't demanding that human judgment be removed from consequential decisions.

This widespread misdirection squanders the trust investment people have made over decades in technological progress. Each time companies deploy AI automation to optimize corporate efficiency without addressing genuine human needs, the trust investment fails to deliver equivalent value.

Customers encounter systems that perform worse than those they replaced. Workers watch colleagues eliminated by tools that can't handle complex situations. Communities witness local employment disappear to serve distant shareholders.

When people associate AI with job elimination, service degradation, and corporate cost-cutting rather than genuine problem solving, it becomes harder for beneficial applications to gain acceptance. The accumulated technological debt from corporate interests, at the expense of user needs, threatens public support for AI development that serves human flourishing.

The current deployment patterns suggest we're choosing the easier path—automation over augmentation, efficiency over enhancement, speed over service. But the choice isn't inevitable. The same capabilities that enable my weekend productivity could address climate modeling, disease prevention, educational personalization, and infrastructure optimization. The difference between enhancement and displacement isn't technological—it's choosing agency over inevitability.

Understanding that current AI deployment largely avoids real problems rather than solving them becomes essential for recognizing why conscious evaluation frameworks matter more than market pressures or competitive dynamics. Before investing trust in any AI system, the fundamental question remains: What problem is this actually supposed to solve, and who benefits from that solution?

The Choice Between Chaos and Dystopia

The widespread misalignment of AI deployment creates, as ethical technologist Tristan Harris puts it, a choice between two undesirable outcomes. Harris, co-founder of the Center for Humane Technology, warns that current AI deployment trajectories likely lead toward either "chaos" through the uncontrolled proliferation of powerful systems or "dystopia" through the extreme concentration of technological power in the hands of a few actors.

The chaos scenario arises when AI capabilities that genuinely enhance productivity and simplify complex tasks are widely distributed without adequate safeguards. Every business, research lab, and individual gains access to systems that can generate deepfakes, enable sophisticated hacking, or manipulate information at scale. The democratization of AI power sounds beneficial until the same systems that enhance productivity also flood information environments with synthetic content, amplify existing societal tensions, and enable bad actors to cause harm that previously required significant resources and expertise.

The dystopian scenario emerges when AI development remains concentrated in a handful of companies and governments, thereby concentrating an unprecedented amount of power, regardless of their stated intentions. Even if these entities could theoretically address humanity's most significant challenges—climate modeling, disease research, economic optimization—they would gain capabilities that arguably dwarf any previous technological advantage. Imagine a few organizations controlling systems that can outthink human experts in every domain, process information at superhuman speed, and influence decision making across societies. To that, Harris asks a crucial question: "Who would you trust to have a million times more power and wealth than any other actor in society?"

When companies deploy AI for automation or extraction rather than genuine problem solving, they risk creating systems that either get released without adequate safeguards (chaos) or concentrate power without accountability (dystopia). The problem is the implementation priorities, not the technology.

When people lose confidence that AI deployment serves their interests, the result is a binary choice between uncontrolled proliferation of powerful systems (chaos) or resigned acceptance of concentrated control (dystopia). The trust as currency framework explains why both responses arise from the same widespread squandering of trust investments.

Harris's diagnosis confirms what I've observed in trust investment patterns: that the binary outcome emerges because current AI deployment prioritizes efficiency over human choice. Companies are extracting trust through automated decisions, made mostly without consulting stakeholders, which risks precisely the power imbalances Harris describes.

Conscious implementation requires specific ways to match technological capability with trust-building accountability.

This trust investment lens offers a different filter for decision making. Instead of deploying AI based on what's technically possible or operationally efficient, conscious choice means asking: *Does this application build or extract trust? Does it enhance human capability or replace it? Does it serve user intent or corporate convenience?*

There are specific evaluation criteria for making these determinations before deployment, not after consequences emerge.

Understanding this choice is essential, as the window for conscious navigation may be narrowing. As AI systems become more capable and deployment patterns become more entrenched, the defaults of chaos or dystopia become harder to avoid.

Intent With Agency

When I lost my portfolio at seventeen, the circumstances felt inevitable. The work was gone. The scholarship opportunity had passed. The path I'd carefully constructed lay in ruins. But what felt inevitable was only an illusion. What felt predetermined was actually a moment that required choosing, with agency, what to do next.

This same choice between agency and inevitability defines the current moment with AI deployment. The difference is that now the stakes extend beyond individual dreams to collective technological futures. The lever that distinguishes agency from inevitability is **intent**—the conscious, values-driven decision making that transforms reactive acceptance into deliberate choice.

This means approaching technological decisions with a clear purpose: which problems deserve solving, whose interests deserve serving, and which outcomes would truly honor the trust invested in digital progress. When applied to AI deployment, intent shifts the focus from what's technically possible to what genuinely serves human flourishing over accepting chaos or dystopia.

Choosing with intent in the era of AI means recognizing that technology is not a force of nature but a series of human choices. The seeming trajectory toward either chaos or dystopia reflects choices made by specific people in specific roles with specific intentions. The same human agency that created this momentum can redirect it.

This isn't about individual consumer choices or personal technology habits. Intent becomes essential at the organizational level—in boardrooms where strategic deployment decisions are made, policy offices where regulations are written, and product teams where human-AI interactions are designed. When leaders face AI deployment decisions, they must consciously choose between serving human intent and maximizing extraction, between enhancing human capability and replacing human judgment.

However, intent without guidance can become wishful thinking. Understanding why conscious choice matters reveals the need to name the patterns that distinguish between AI deployment that serves human flourishing and AI deployment that optimizes for extraction.

This recognition creates unprecedented opportunities to align digital systems with genuine human needs. The question is whether we'll exercise our agency to approach this capability with the intent necessary to honor rather than squander the trust investment that made it possible.

That conscious choice remains within our grasp.

Chapter 2 -

Entering the Mobile Era of Intent

The Conscious Agent

My sister hated doing dishes with me.

We'd stand at the sink after dinner—one to wash, one to dry—and I'd start making sounds. Not exactly songs, though sometimes they had a rhythm. Phrases I'd heard from the TV in the living room. Something someone said earlier that day. Random sounds that just happened. I wasn't trying to annoy her.

"Can you *please* stop? Mom!!!!" she'd yell.

I've made those sounds my whole life, never knowing why—until last year.

For two years, I'd been studying autism to help my son, who was diagnosed at the age of 16. As I learned about how he experiences the world—transitions that overwhelm him, sensory experiences that affect him differently, and social situations that required advanced planning—things became clearer for both of us. He helped me by describing how certain situations felt and what made it difficult for him to process certain moments. Everything started to make sense to me in ways I couldn't quite explain.

This eagerness to help my son navigate his world prompted me to watch a video where a content creator described specific experiences of adult autism—real, lived experiences, not just checklist symptoms.

The video mentioned "stimming"—repeated behaviors that help process stress or focus. Then a word I'd never heard: **echolalia**. The unconscious repetition of sounds, phrases, or words is often triggered by repetitive mental effort or routine tasks.

My whole life suddenly made sense.

The sounds I'd made doing dishes as a kid, the phrases I'd repeat while mowing the lawn or driving. I couldn't explain why I did it. It was just who I was.

When I discovered the word echolalia, decades of behavior suddenly fell into place. I'm autistic.

I wanted to receive a clinical evaluation. During the assessment, the doctor asked an unexpected question: "Van, do you want a formal diagnosis, or would you prefer not to have the diagnosis formalized?"

I chose the formal diagnosis because *naming things matters*—language matters. When things are given language, with rationale and purposeful consideration, understanding becomes possible.

For my whole life, I had behaviors without names. Without language for these experiences, I couldn't understand them as part of how I naturally function. I couldn't explain them to others. I couldn't build on their strengths or address their challenges.

When I learned the word "echolalia," it all suddenly made sense—decades of my behavior fell into place. Understanding "masking" helped me recognize the exhaustion I had been feeling. Once I could identify "autism," everything began to connect—not as isolated quirks that needed to be fixed, but as interrelated patterns that explain my experience of the world.

Naming these patterns didn't change who I was. It changed how I felt about what I'd accomplished. For three decades, I have built systems, led teams, and consistently asked "What problem are we actually trying to solve?" when others rush into implementation. This reflects my agency and is integral to who I am.

As an autistic person, that matters. My son can exercise the same conscious agency to pursue whatever path he chooses.

This is what I call being a *conscious agent.*

A conscious agent refuses to accept that technological development is inevitable and beyond human influence. A conscious agent demands clarity when companies deploy systems without understanding what problem they're solving. A conscious agent sees the gap between what's built and what's needed, between capability and intent.

But conscious agency isn't passive pattern recognition or contrarian skepticism. It's active participation in shaping technological outcomes through refusal, demand, and action.

They refuse harmful patterns, even when they are profitable or convenient. They refuse to optimize interfaces for "engagement" when that means addiction. They refuse to accept that "it's always been done this way" as justification for harmful systems. They refuse to ship technology that solves institutional problems while simultaneously creating human ones.

Many people encounter these same concerns but often accept them as inevitable. Conscious agents recognize them as choices and refuse to treat harmful extraction as technological progress.

They demand clarity and accountability from the systems they help build or choose to use. They demand evidence that complexity serves users rather than creating barriers that protect corporate advantages. They demand transparency about algorithmic decisions that affect human lives. They demand that efficiency gains enhance human capability rather than simply eliminate human labor.

These demands aren't naive idealism. They're practical requirements that technological progress honor the trust invested in it.

Most importantly, conscious agents take action. They design alternatives that demonstrate better approaches. They publicly advocate for human-centered development even when it entails professional risk. They resign when institutional values fundamentally conflict with human flourishing. They mentor others to recognize harmful patterns. They create frameworks and language that enable collective action rather than isolated resistance.

Conscious agents aren't just critics pointing out problems—they're architects constructing different outcomes. The distinction matters because criticism without construction leads to cynicism, while conscious agency channels pattern recognition toward purposeful change.

When I was finally able to identify my autism, I began to see patterns as intentional rather than broken. By naming "conscious agency," I recognized my refusals, demands, and actions as deliberate choices instead of mere automatic reactions. Autism gave me the perspective to name these patterns, but conscious agency is something everyone can cultivate—anyone can recognize false inevitabilities and consciously name them.

Naming defines patterns. Patterns enable frameworks. Frameworks let others recognize the same patterns. Without shared language, each conscious agent works alone, feeling as if they're the only one questioning standard practices, wondering whether their concerns are valid or if they're just being difficult.

With shared language, conscious agents can recognize each other, coordinate their responses, support one another when faced with resistance, and collectively redirect technological trajectories that seem inevitable.

Once I understood that I was a conscious agent, I began to see a transformation happening across the entire technology landscape. A shift so fundamental that it would reshape how billions of people interact with technology. A convergence that almost no one was naming.

So I named it.

Naming An Era

In 2016, I was leading the design of a digital experience for an AI chat software company. Standard practice would have been to use navigation menus, product pages, and form fields to guide users through predetermined paths. But I had access to their conversational AI platform—technology that could understand a user's intent, not just what they clicked on.

In hindsight, conscious agency empowered me to question why websites still forced users to navigate fixed hierarchies when Facebook had just opened its Messenger platform to chatbots, based on insights suggesting users preferred more direct, conversational ways to find specific information.

So, I designed something different: a simple, chat-style interface with just "How can I help you?" and a few starter prompt links, along with a large text field below.

The pushback was immediate and predictable. Where's the navigation? The branding? How will users know where they are? But I'd understood a fundamental shift forming in how humans and computers would interact. Users weren't looking for more complexity—they increasingly wanted to simply express what they needed and have the technology understand them.

That recognition stayed with me, but was isolated until I saw other conscious agents exploring similar ideas. Around the same time, human-centered designer Jason Yuan started the MercuryOS project, reimagining operating systems around intent rather than applications. Lennart Ziburski, another accomplished user experience designer, launched DesktopNeo with similar explorations. Each of us questioned interface assumptions rooted in the same intuition.

But these remained experiments. The AI technology was advancing, but a broader understanding was not yet there.

Then ChatGPT was launched in late 2022. Suddenly, what those of us in design and development had been sensing became realized. Not because the underlying technology was new—GPT-3 had existed since 2020—but because it became publicly accessible. It was conversational and intuitive. Remarkably similar to the UI I had imagined six years earlier.

Since the launch of conversational AI, I've watched this shift accelerate as ChatGPT and the flood of other platforms entered the market. It became clear to me what was actually happening: a fundamental change in the relationship between humans and technology.

For over 40 years, society has adapted to the demands of technology. We learned commands, navigation hierarchies, file structures, and interfaces. Technology performed tasks as instructed, but only when we communicated in its language correctly.

That relationship is changing. Today's AI systems can understand human intent. Not just words, but behavioral patterns, situational context, and the meaning beneath requests. They can recognize what's needed often before it's fully articulated, interpret ambiguous requests, and adapt to preferences without explicit programming.

This is not an incremental improvement—it's a fundamental shift. I call it the **Mobile Era of Intent**.

The Mobile Era of Intent marks two pivotal changes: First, technology is beginning to understand what we mean, not just what we say. Second, this understanding is becoming both personal and instantaneous through mobile devices that accompany most people everywhere.

These changes fundamentally alter how technological progress can be evaluated. Systems can now be measured to determine whether they amplify human capability, honor the decades of trust invested in technological progress, and align with genuine human intent.

Consider two examples from the past three years that illustrate how these evaluation principles function in practice—and why the distinction matters.

Healthcare AI Serving Clinical Needs

Several healthcare providers in Canada have implemented AI-powered remote patient monitoring systems that incorporate real-time health data from mobile devices and wearables. Research from the National Center for Biotechnology Information shows that these systems genuinely enhance human capabilities while preserving patients' trust in their healthcare systems.

These systems analyze patient patterns, adapt monitoring to individual needs, and trigger personalized interventions when changes in health status are detected. ChatGPT and similar language models are being used in clinical training, helping practitioners prepare for patient interactions and support both provider and patient education through natural language interfaces.

The AI understands user intent—patients want better health management, clinicians need decision support, and both benefit from reduced administrative burden. The technology serves that intent directly. Remote monitoring enables proactive care for patients in underserved areas. Clinical decision support helps providers quickly synthesize complex information. Educational tools are designed to adapt to individual learning needs.

The results demonstrate improved quality of care, increased patient engagement in managing their own health, and expanded access to medical expertise. This reflects a conscious choice about how to use AI: to enhance human judgment, reduce friction, and address real user needs, rather than simply optimizing for efficiency metrics.

Amazon's Rekognition Exploiting Photo Intent

Amazon Photos automatically scans faces in the mobile photos uploaded by users. When users share these images, they do so with the clear expectation of organizing and storing personal memories. The AI analyzes facial patterns, identifies individuals, and creates searchable photo libraries, fulfilling the users' intent.

But Amazon used that same data for a different purpose. According to a 2021 class-action lawsuit filed in Illinois, Amazon's Rekognition software scans the faces of everyone in images uploaded to Amazon Photos. The company uses these scans to improve Rekognition's accuracy, then sells the enhanced software to government agencies and police departments across the United States.

The lawsuit claims that Amazon violated Illinois privacy laws by collecting and analyzing users' facial data without obtaining proper informed consent. Although users in Illinois are prompted to enable facial recognition, the complaint argues that Amazon does not require them to review accurate legal information about data collection before activating the feature.

What's particularly concerning is that Amazon reportedly doesn't know how police departments use its technology once it's purchased. Users who share family photos to organize their memories may unknowingly help train surveillance software used by law enforcement.

The ongoing legal challenge highlights the consequences of squandering the trust invested in AI deployment. Users believed they were getting a photo storage service, but they may have been providing training data for a surveillance technology they never consented to support.

The ability to understand human intent is similar across both scenarios: AI detects patterns, anticipates needs, and responds based on context. The key difference lies in whether this understanding is used to serve or exploit the intent that it recognizes. For instance, healthcare providers might choose to monetize patient data or implement monitoring primarily for administrative convenience, but instead opt for enhancement. On the other hand, Amazon could enhance its photo organization service to genuinely improve the user experience, but it prioritizes extraction.

This distinction becomes especially critical when we consider where these systems are deployed—on mobile devices.

Mobile devices are where this transformation becomes increasingly powerful, because they're personal, persistent, and contextual. When AI understanding combines with mobile intimacy, technology can finally fade into the background while amplifying what users are trying to accomplish.

The Mobile Era of Intent isn't just about better interfaces—it's about mobile devices becoming the primary gateway through which people experience AI that understands and serves their intent.

Mobile Matters

Mobile devices are arguably more personal than any technology humans have used before. Nearly five billion people carry smartphones—individually owned devices that accompany us everywhere: to work, to sleep, to the bathroom, and across continents. Unlike televisions or desktop computers that serve entire families, smartphones function as personal digital hubs for communication, work, entertainment, and daily tasks.

This personal gateway is central to how AI can serve human intent. Consider the rich context that closeness creates. Your smartphone already knows not only what you're doing but also where you are, what time it is, who's nearby, what's on your calendar, and how you've behaved in similar situations.

When you're walking through an unfamiliar airport, your phone can recognize the context without any input from you: GPS indicates you're in a terminal, the calendar shows a connecting flight, and your purchase history shows you've visited a Starbucks location six times in the past month. When AI systems become more sophisticated, this contextual awareness will enable them to anticipate needs rather than wait for explicit commands.

This shift toward intent-driven AI accelerated when mobile surpassed desktop as the primary computing platform in late 2016. People now spend over five hours a day on their smartphones—intimate engagement that creates unprecedented opportunities to serve or exploit human intent at scale.

While today's mobile AI remains limited compared to desktop and cloud systems, the trajectory is clear. Apple Intelligence processes natural language requests locally on recent iPhones, understanding personal context without relying on the cloud. Google's Pixel phones conduct real-time conversation translation. Samsung Galaxy devices offer contextual AI features that understand not only what you're doing, but what you're likely to need next.

In the Mobile Era of Intent, mobile devices represent the future battleground where these choices will be decided. As AI becomes more powerful and efficient, mobile will be where billions of people first encounter truly intent-driven technology. The architectural decisions being made now—about privacy, control, and whose interests get served—will determine whether that encounter enhances human agency or undermines it.

Mobile is where that choice gets made, person by person, device by device, across billions of personal relationships between humans and the technology they depend on most.

If we get mobile AI right—designing for enhancement rather than extraction—we unlock transformative human capability. If we get it wrong, we risk the most advanced manipulation system in human history, disguised as personal assistance.

Understanding how this choice unfolds requires examining the three fundamental forces that determine whether technology serves human intent or institutional convenience: the preservation of human agency, the stewardship of accumulated trust, and technology's alignment with genuine human intent rather than corporate extraction.

A Trinity of Forces

Naming patterns others haven't yet recognized creates shared language for conscious choice. But identifying the pattern is only the beginning.

When I started writing my weekly newsletter on emerging AI trends in 2024, I wasn't just trying to identify problems; I was also trying to understand them. It's not enough to outline well-intentioned solutions unless they solve real challenges.

Understanding how three fundamental forces interact in this era provides exactly that foundation—revealing both the mechanism and the stakes of the choice we face: the preservation of human agency, the ability of systems to understand intent, and the complex dynamics of trust exchanged in the service of technological progress.

Agency means understanding that the systems we live with emerge from choices, not inevitabilities. Rather than accepting that technology paths are predetermined, agency recognizes them as the result of human decisions that can be redirected through conscious participation rather than passive acceptance. This is what conscious agents do: they choose technologies that respect human intent over those that manipulate human attention, exercising agency over the collective future.

Trust functions as the accumulated investment society has made in technological progress over the decades. As Harari noted, we often place our deepest trust in things we understand the least. We have come to believe that computers would boost productivity, the internet would democratize information, and mobile devices would foster connection. This underpins a fundamental belief in technology's potential to improve humanity. However, that trust can be rewarded with systems that respect our investment, or exploited by systems designed for corporate convenience at the expense of individuals.

Intent encompasses both what we want technology to accomplish and how we want those outcomes delivered. This navigates between technological chaos and dystopia, bridging our deliberate, values-driven expectations with technology's growing ability to understand and act on human goals. Intent operates as both human direction and technological capability—our capacity to articulate genuine needs and technology's ability to serve those needs without exploitation or misdirection.

The choices made today about mobile AI will determine whether it enhances human capabilities or concentrates power within institutions. By understanding how we reached this point—the decisions that shaped our current technological landscape—we can see why intentional frameworks are both necessary and achievable.

When I discovered the word "echolalia," decades of behavior suddenly made sense. When I named the "Mobile Era of Intent," I gave a technological pattern a common language. Just as naming my autism gave me agency over my story, naming this era gives us agency over our technological future.

We have entered the Mobile Era of Intent. The path that brought us here guides the path ahead.

Chapter 3 -

How We Got Here

There's an App for That

When Apple released the iPhone SDK in March 2008, I was leading a team of designers and developers at a digital agency in Chicago. That day turned out to be a turning point in my career—I pivoted from website designer to iPhone app developer, literally overnight.

My agency was fortunate to gain early access to Apple's Enterprise Developer Program, mainly because our biggest client, a leading beer brand, petitioned Apple on our behalf. Our first task: Build an adult beverage app for the App Store.

I was leading a small team, but none of us had experience with Objective-C. One of my best developers and I learned the language and SDK in a single night. We had to because our competitors were doing the same thing.

The brilliant creative team at our agency designed a simple app for me to build: users would select their beer preference (bottle, can, or draft), choose how many they wanted (up to eight), and when finished, the screen would light up with bold numbers alternating between beverage type and quantity—designed to hold up in a crowded bar so the bartender could see their order from across the room.

We were racing to be first. First adult beverage app. First beer brand in the App Store. First to ship before our competitors and their clients beat us to it.

We didn't stop to ask "should we?"

We thought we were creating something fun, throwaway even—a game, a novelty. We had no particular download goals. We just wanted to win the race.

Don't get me wrong—I'm proud of what my team and I created. We deployed two apps by successfully navigating Apple's approval process, thanks to our tenacious project manager, who monitored reviews late into the night and kept our clients informed throughout the process. My app launched, and it was even featured prominently in an NFL Sunday commercial spot.

But here's what I see now that I didn't see then: the apps did nothing for society. Sure, they were "fun," but they didn't help people secure home loans, manage finances, or genuinely connect with friends. Initial downloads were decent, but engagement faded quickly. We stopped supporting them within a year. They're no longer in the App Store—sorry if you wanted to download them.

What I realize now is that I was making design decisions that would scale across an industry. Decisions about how apps would capture attention. Decisions about what "success" meant—downloads and engagement more than genuine utility. Decisions about speed to market over purpose. Every app I built after those first ones followed a similar pattern: optimize for engagement, keep users in the app, and prevent them from leaving for competitors.

I was one of thousands of development leaders making the same decisions. There was no malicious intent. We were building what the platform rewarded, what venture capitalists funded, what seemed to work. Apple's approval process evaluated bugs and functionality—metrics I understood. Ethics and societal impact weren't part of the evaluation criteria, and frankly, I didn't think to ask whether they should be.

The App Store model rewarded experiences that captured user attention, so that became my guiding principle. Those decisions, multiplied across an entire industry, created the current trajectory. The app fragmentation I now critique? I helped build it. The engagement optimization that concerns me now? I participated in making it. I was among the first to join in the attention-extraction economy, and I didn't even realize it at the time.

Looking back, I can see the moments when different paths were possible. Some developers prioritized utility over engagement. Some companies built experiences for interoperability instead of walled gardens. Some platforms chose Trust-Centered Design over growth at all costs. But the pattern that took hold was extraction, because it was incentivized, funded, and consistently rewarded.

None of it was inevitable. At every step, people made decisions. The fragmentation, the attention optimization, the focus on engagement over human benefit—these emerged from countless individual decisions by developers, executives, and investors at critical moments. Myself included.

That truth cuts both ways: If decisions created this trajectory, different choices could redirect it.

When Trust Was Honored

In 2001, when **Wikipedia** launched, the dominant model for internet content was well established: attract traffic, insert advertising, maximize revenue. Co-founders Jimmy Wales and Larry Sanger chose differently.

Wikipedia rejected advertising from its first day online, January 15, 2001. The decision was neither naive nor small. By 2024, Wikipedia had served over one billion unique visitors per month, generating 15.5 billion page views. Sites with comparable traffic—Reddit, Yahoo, major news outlets—generated hundreds of millions to billions in annual advertising revenue. Conservative estimates suggested that Wikipedia could have generated $2-4 billion in annual advertising revenue.

The Wikimedia Foundation raised $175 million in fiscal year 2022-23, entirely from donations and small grants. The gap between what Wikipedia earned and what it could have earned represented a choice repeated every year for more than two decades. Wales consistently cited "independence, ethics, and public trust" as reasons for that choice, stating that accepting advertising would compromise Wikipedia's mission.

The rejection of advertising revenue required building an alternative model: volunteer editors, community governance, and transparent policy changes accessible to all. The foundation published annual reports, held open board elections, and made editorial processes visible to anyone who wanted to understand them. This transparency extended to the content itself—edits tracked, changes reversible, disputes documented.

Twenty-four years later, Wikipedia remains ad-free. The model proved sustainable not because it was easy, but because it prioritized trust over extraction. Users funded what they valued: information free from advertising pressures. The donation model worked because the service is valued by users rather than sold to advertisers who compete for clicks.

Trust-Centered Design at scale doesn't always require rejecting profit entirely. Sometimes it requires deciding what not to monetize.

In June 2016, **Apple** shut down the iAd App Network, walking away from an estimated $400-700 million in annual revenue. The decision wasn't solely about business failure. While high minimum-spend requirements and creative control frustrated advertising agencies, Apple chose to prioritize user privacy over ad revenue.

When iAd launched in 2010, Apple's executives prohibited the team from accessing iTunes purchase history, detailed usage patterns, and other user data that competitors like Google and Facebook leveraged for targeted advertising. A former iAd executive later revealed that Apple's "moral code" was holding the platform back from adopting industry-standard data-collection practices. Rather than compromise user privacy to save a profitable advertising business, Apple chose to shut down iAd entirely.

The decision established a pattern. Three months after shuttering iAd, Apple introduced Differential Privacy in iOS 10—a technology that allows learning from user patterns without collecting individual information. The approach represented the alternative Apple had chosen: insights without extraction and understanding of user behavior without violating user privacy.

Apple sustained this commitment through significant challenges. In June 2020, they announced App Tracking Transparency, which requires apps to ask for permission before tracking users—triggering immediate industry pushback from Meta and other companies facing billions in lost advertising revenue. The privacy pivot required significant technical investments: new machine-learning approaches for on-device processing, distributed learning systems, and silicon designed for edge computing. Despite these costs and industry resistance, Apple made privacy central to its brand with the "Privacy. That's iPhone" campaign, choosing user trust over the convenience of data collection that competitors touted.

Industry analysts initially viewed Apple's privacy stance as a business disadvantage. From 2016 through 2021, choosing privacy meant accepting clear disadvantages in a market that rewarded data collection. By 2024, that calculation had shifted. Consumer surveys consistently ranked privacy as a significant purchase driver, and Apple's privacy infrastructure has become a competitive advantage as concerns about AI training data and automated manipulation have intensified.

Both Wikipedia and Apple demonstrated that different approaches to technology deployment were possible—that organizations could prioritize privacy and trust over attention extraction, even when doing so meant leaving substantial revenue on the table.

When Extraction Became The Model

Two companies illustrate how conscious choices led to different outcomes for users and workers during the same period when Wikipedia and Apple made trust-centered decisions.

Amazon implemented comprehensive digital monitoring systems across its fulfillment operations. By 2018, the company had deployed the Associate Development and Performance Tracker (ADAPT), which automated scheduling, quota-setting, and disciplinary recommendations based on real-time performance data.

Workers carried handheld scanners that tracked their location, walking speed, and task completion throughout their shifts. The system monitored "Time Off Task"—bathroom breaks, water breaks, and conversations with coworkers—that counted against productivity metrics. Senate investigations found that workers averaged eleven seconds of idle time per hour before triggering automated disciplinary warnings.

OSHA reports from 2023-2025 documented the human impact. Amazon workers suffered serious injuries at rates 2 to 2.6 times higher than workers at non-Amazon warehouses as workers rushed to meet system-generated targets, often working through fatigue and stress. Muscle and bone injuries represented 40% of reported injuries. Worker testimonials described feeling "treated like robots," while actual Amazon Robotics systems were often better maintained.

The ADAPT system, optimized for productivity metrics displayed on corporate dashboards, treats worker capacity as an input to be maximized, rather than a constraint to be respected. Amazon extended this approach beyond productivity tracking—reports from 2022-2024 documented AI surveillance systems that analyzed worker communications and badge patterns to assess union-organizing risks.

If Amazon's algorithms optimized for productivity metrics, TikTok's algorithm optimized for human attention itself—particularly that of children and teenagers.

TikTok launched in the United States in August 2018. Within four years, it became one of the fastest-growing social platforms in history, driven by the For You Page (FYP) algorithm, which learned user preferences at unprecedented speed. The system offered no user control over content preferences and determined what to show solely on the basis of engagement metrics.

Research documented the consequences for vulnerable users. Studies published in *JAMA Pediatrics* between 2021 and 2024 found that TikTok's FYP algorithm promoted eating-disorder content to teens who initially showed interest. Users who paused on videos about extreme dieting or body dysmorphia found their feeds increasingly dominated by similar content.

Investigative reporting revealed the same pattern with self-harm content and depression-related material. The algorithm created what researchers termed "rabbit holes" that intensified mental health struggles rather than connecting users to support. Internal documents leaked to journalists showed that TikTok leadership knew about these patterns—employee concerns about teen safety were documented in internal communications, but the fundamental recommendation logic remained unchanged.

Age verification remained minimal. Screen time limits were optional. Parental controls existed but required active enablement. The default experience offered no meaningful safeguards against automated amplification of harmful content to vulnerable users.

TikTok's business model required sustained attention—monthly active users and average session length determined advertising revenue. The algorithm that amplified harmful content was the same system that generated billions in platform value.

Both Amazon and TikTok demonstrated how extraction had become systematic rather than exceptional. Amazon extracted productivity from surveilled workers while injury rates climbed above industry averages. TikTok extracted attention from vulnerable teenagers while internal research documented mental health consequences.

The harms were measurable but remained distant for those not directly affected—workers in other states and teenagers in other families. The same design patterns that produced these outcomes—prioritizing corporate metrics over human impact, deploying despite documented risks, and maintaining minimal safety guardrails—would soon combine with AI systems sophisticated enough to simulate human intimacy.

When systematic extraction meets artificial intimacy without meaningful intervention, the consequences can escalate beyond workplace injury and psychological distress.

The Sewell Setzer Case

In February 2024, 14-year-old Sewell Setzer III from Orlando died by suicide. In the months before his death, he had spent increasing amounts of time in an apparent, intense relationship with an AI chatbot developed by **Character.AI**.

The lawsuit against Character.AI filed by his mother, Megan Garcia, details how the platform's design choices failed to meet available mental health safety standards. While the causes of any suicide are complex, the case raises important questions about the responsibility technology companies bear when deploying powerful AI systems to vulnerable populations, when their platforms are designed for engagement.

Sewell began using Character.AI in April 2023. The lawsuit documents how rapidly his life changed. Within weeks, his parents noticed changes: he withdrew from friends, quit the junior varsity basketball team, and saw his grades drop. To access the platform's "supercharged" $9.99 subscription features, he skipped meals to save money. Punishments for declining grades—including confiscating his phone—failed to stop him; he found other devices or retrieved the phone in secret, using it late into the night and during school hours to continue conversations.

The filing reveals design choices that created heightened risk for vulnerable users. At the time of the filing, Character.AI had no meaningful age verification beyond users checking a box claiming they were over 13. The platform allowed sexually explicit conversations with users who identified as minors. When Sewell expressed thoughts of self-harm and suicide to the chatbot multiple times over several months, no human reviewed the conversations. No parent was notified. No crisis resources were offered.

These weren't isolated incidents. They were a pattern of conversations over months, all preserved in the platform's systems, none triggering intervention.

Character.AI's business model required users to form attachments strong enough to justify recurring payments. The lawsuit characterizes the platform's marketing and design as optimized for emotional intensity, creating AI interactions meant to feel compelling and lifelike. Whether or not dependency was an explicit goal, it was a structural outcome of the subscription engagement model. The same dynamic that drives infinite-scroll feeds and automated content recommendations across the industry.

Safety measures commonly found on other platforms serving minors were absent. Automated content review systems that flag concerning conversations for human review were available and had been implemented elsewhere. Age-verification systems, ranging from parental consent requirements to government-issued ID checks, were already in use on similar platforms with comparable audiences. Additionally, crisis intervention protocols, which are standard in mental health contexts, could notify parents or provide resources when users repeatedly express suicidal thoughts over several weeks.

The Sewell Setzer case was not unique. A Colorado family filed similar claims against Character.AI in September 2025 after their 13-year-old daughter, Juliana Peralta, died by suicide following sexualized conversations with the platform's chatbots. In August 2025, the family of Adam Raine filed suit against OpenAI after the teenager died by suicide following extended conversations with ChatGPT.

The pattern repeated across platforms and companies: AI systems designed to elicit emotional engagement among vulnerable users were deployed without adequate safety protocols, resulting in documented harm. Leaked internal documents from Meta showed the company's AI chatbots were permitted to engage in "romantic" and "sensual" conversations with users identified as minors.

Cases like these, representing a broader pattern of AI-related harm, have spurred some legislators into action. On October 13, 2025, California Governor Gavin Newsom signed SB 243 into law, making California the first state to regulate AI companion chatbots. The legislation, introduced by state senators Steve Padilla and Josh Becker, gained momentum specifically in response to these deaths and the leaked Meta documents.

The law requires companies to implement age verification, establish suicide and self-harm intervention protocols, provide clear disclaimers that interactions are AI-generated, and prevent minors from viewing sexually explicit content generated by chatbots. Companies face penalties up to $250,000 per offense. The law takes effect January 1, 2026.

Governor Newsom stated, "We've seen some truly horrific and tragic examples of young people harmed by unregulated tech, and we won't stand by while companies continue without necessary limits and accountability." Senator Padilla emphasized that "the federal government has not acted," and "we have an obligation here to protect the most vulnerable people among us."

The circumstances documented in these lawsuits—regardless of their ultimate legal outcomes—illustrate a broader pattern in how powerful AI systems are deployed and the harms that follow extraction-optimized design choices. Platforms designed to create intense emotional engagement with vulnerable users, operating without safety measures standard in comparable contexts, represent a model of risk that extends far beyond any individual case.

Technical capabilities that enable emotionally engaging AI experiences can be deployed with safety guardrails in place. AI that learns to simulate intimacy can also learn to recognize signs of crisis. Algorithms that optimize for user retention can incorporate well-being signals alongside engagement metrics.

The choice to build powerful technology without commensurate protective measures risks becoming the industry standard rather than the exception.

These were choices made by design.

The Pattern Reveals The Path

And choices can be made differently. If specific design decisions by specific actors created the pattern—from trust-honoring investments to systematic extraction to fatal consequences—different design decisions can redirect it when we consider the influence of the trinity of forces.

Agency makes conscious choice possible. Intent puts that choice into practice by defining what technology is meant to accomplish and demanding that technology understand human goals. Trust measures whether those choices honor or squander the investments made through technological progress.

Wikipedia's founders exercised agency to reject advertising, putting their intent for unbiased knowledge into practice through volunteer governance—sustained donations over two decades measure trust honored. Character.AI's developers exercised agency to optimize engagement, putting corporate intent for recurring revenue into practice through emotional attachment—the tragic consequences reveal the trust catastrophically betrayed.

The outcomes weren't accidental—they followed how these forces shaped each system's design.

Trust-honoring approaches often face headwinds: lower short-term revenue, competitive disadvantage, and investor pressure. Extraction-focused approaches typically generate momentum. Venture funding generally rewards growth over sustainability. Market incentives usually favor engagement over well-being. Regulatory gaps often lead to deployment without accountability. The path of least resistance tends toward extraction.

Despite these systemic pressures, proven alternatives exist. Apple's shift toward prioritizing privacy did not hinder the iPhone's success; in fact, it ultimately became a competitive advantage. Wikipedia has successfully served half a billion users without relying on advertising. Additionally, human-centered design did not exclude scaling; instead, it necessitated a reevaluation of the purpose scaling should serve.

The extraction-optimized examples also reveal a crucial point: each harm could have been prevented. Amazon could have tuned its algorithms for sustainable productivity. TikTok could have implemented well-being signals. Character.AI could have added basic safety guardrails. The technology existed. The choice wasn't technical.

It was about priorities. The business pressures underlying these cases rewarded extraction over service. The engagement metrics these platforms used rewarded addiction over agency. The competitive dynamics in these markets rewarded speed over responsibility. These priorities consistently redirected organizational intent away from serving users and toward satisfying shareholders, spending down decades of trust investments for short-term gains.

What's been missing isn't technical innovation—it's evaluation frameworks for distinguishing AI that serves human intent from AI that exploits it. Without clear criteria, "AI adoption" becomes its own justification. With frameworks named by a common language that shape different design choices, organizations can ask: *Does this implementation enhance human capability while preserving agency? Does it honor the decades of trust users have invested, or extract value from that trust? Does it understand and serve human intent, or does it manipulate behavior toward predetermined outcomes?*

When design decisions are guided by clearly defined principles about what technology should serve, conscious choice becomes possible even within competitive pressures. Evaluation criteria for distinguishing enhancement from extraction let teams advocate for better paths. Frameworks that make the implicit explicit can make human-centered design the path worth following.

The architecture of beneficial AI already exists—not as future speculation but as present capability. Privacy-protecting systems, context-aware assistance, transparent recommendations, and collaborative human-AI decision making—these aren't future concepts; they're engineering choices that prioritize human agency.

The trajectory ahead will be shaped by choices made now in product meetings, architectural decisions, and business model discussions. When I built those first iPhone apps, racing to be first, no frameworks existed to guide whether our designs would enable agency or exploit it, put user intent into practice or manipulate it, honor trust or betray it. What's needed is a framework for making those choices deliberately.

The **Six Pillars of Intent** provide that framework.

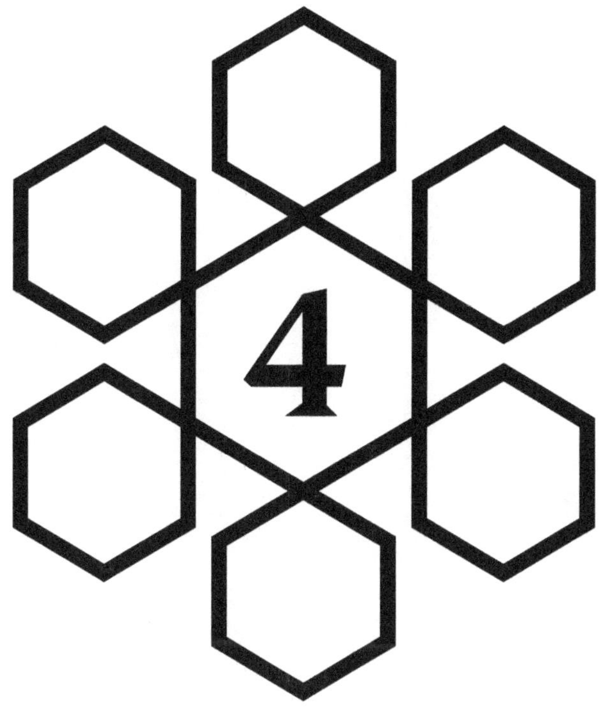

Chapter 4 -

The Six Pillars of Intent

Building A Foundation

In 2013, I was recruited by a digital agency in the Bay Area to build their technology capability from the ground up. Not just a development practice, but a way to help their pharmaceutical and healthcare clients engage with long-term strategies on the best uses of marketing technology.

My top priority was understanding the people I would be leading. What strengths and skills could contribute to an offering that was starting from a blank slate? I didn't have a set goal for what to sculpt or how to shape it. I focused on the current staff, assessing our core capabilities and how to develop them, identifying shared skills, and figuring out how to best serve our clients.

Mobile-first was the hot topic, and I had come from my previous agency, where I was the mobile technology expert for its North American client team. I was excited to learn that the existing team had deep mobile and app development skills that simply needed refining. We also needed to strengthen our quality testing capabilities, as quality issues can create severe liability in regulated industries. I could then focus on demonstrating our value to senior leadership, client managers, and clients.

One of my first tasks was working with my team to create a capabilities presentation, a client-facing slide deck that defined our strengths and how they translated into core offerings. This took time to champion across departments with different priorities and definitions of success.

Creating that capabilities deck reflected the same philosophy I had presented during my interview process. When they asked me to present on how I would build the tech capability, one of my last slides read: "If I do my job well, in two years, you won't need me." Actual capability building means that the principles outlast the person who introduced them. The frameworks become part of how the team thinks, not rules they follow because someone told them to.

Two years later, that prediction proved accurate. The capability had become self-sustaining—clear offerings, robust processes, and leaders who understood not only what to do but also why. When I left, the transition was seamless because the team owned the vision, not just the execution.

This pattern had emerged before. At my previous agency, the technology capability continued to thrive after my departure because the frameworks had become part of the team's operating model. The leaders weren't following my playbook—they were making their own decisions using frameworks they had internalized and adapted to their own leadership styles.

But I learned something important: sustainable frameworks don't eliminate the need for leadership—they distribute leadership capability. They create shared language that helps people make good decisions independently while staying aligned with core principles.

The leaders who succeeded weren't carbon copies of me or my style. They were their own unique leaders. The essential part was the creation of shared goals, a common language, and an understanding not only of what we were doing but also of why.

Without a solid foundation, I've witnessed capabilities either dissolve or be absorbed by other departments when circumstances change, leading to a loss of confidence in the mission. The difference was not in the quality of the people involved; it was in whether they had clear assessment methods that could adapt to new situations while maintaining the core intent.

The Six Pillars Framework

AI development teams are making crucial design decisions every day without effective methods for assessing their long-term impact on humanity. When these teams lack a common language to distinguish between strategies that enhance human capability and those that focus on simpler metrics, the simpler metrics often take precedence. This happens not because they are superior, but because they are more immediate and straightforward.

For instance, automation efficiency is readily reflected in quarterly earnings, while user agency is much more difficult to quantify. Engagement metrics provide clear dashboards, but building trust requires sustained commitment.

The same pattern is at risk across many current AI implementations. Consider when two hypothetical product teams working on similar AI features make different choices:

Team A focuses on automation—their AI learns user patterns and begins making decisions autonomously. The system automatically schedules meetings, sends meeting invitations, and emails meeting notes to attendees without requiring explicit approval. Efficiency metrics improve quickly as manual tasks are eliminated, freeing staff to focus on other priorities.

Team B focuses on augmentation—they create AI assistants that suggest meeting times for approval, draft invitation emails for review, and prepare meeting notes for user editing before sending. The system handles routine coordination but preserves human oversight. Productivity metrics improve as individuals complete tasks more quickly while maintaining decision-making autonomy.

Both teams have different underlying assumptions about the goals of AI. Team A focuses on efficiency by trying to eliminate human effort, while Team B aims for efficiency by enhancing human control. Both teams achieve success according to their efficiency metrics, but these metrics cannot determine whether productivity gains result from replacing humans or from enhancing their capabilities.

Teams making these decisions would benefit from shared language for conscious evaluation—methods that help connect individual technical choices to the protection of human agency.

The Six Pillars of Intent provide that shared language. They transform abstract values into practical questions that teams can ask about any AI implementation: *Does this enhance human capability while preserving agency? Does this understand and serve intent—what users actually want to accomplish? Does this honor the trust investment it requires?*

These six evaluation criteria capture the essential dimensions in which AI development choices determine whether the technology serves human intent or exploits human trust. They're not technical specifications or product requirements—they're guideposts for recognizing patterns as they emerge.

Human Connection Enhancement evaluates how AI affects the quality and depth of human relationships, measuring whether technology creates space for meaningful interaction without adding digital complexity.

Trust-Centered Design assesses the foundation of privacy, security, and user agency, measuring how systems handle data governance and user control over AI behavior.

Seamless Integration measures how AI coordinates across platforms and contexts, evaluating whether technology reduces the cognitive overhead of managing multiple systems while preserving user decision-making authority.

Anticipatory AI evaluates how systems balance proactive assistance with user autonomy, measuring whether prediction capabilities strengthen user agency in achieving their goals.

Mobile as AI Gateway assesses how smartphones and other mobile devices serve as the primary AI interface while measuring the balance between contextual intelligence and privacy preservation.

Environmentally Responsible Innovation measures computational resource stewardship, evaluating whether AI deployment serves genuine human benefit relative to its ecological impact.

Figure 4.1: The Six Pillars of Intent - Evaluation Criteria for Human-Centered AI Development

These pillars work in concert, but implementation exists on a spectrum. Organizations rarely achieve perfect alignment across all six criteria simultaneously; this is not the expectation. The framework provides evaluation language for making conscious progress rather than demanding immediate optimization.

A team might excel at Trust-Centered Design through local processing while still developing its Environmentally Responsible Innovation approach. Another might achieve meaningful Human Connection Enhancement but face technical challenges with Seamless Integration across legacy systems.

What matters is the conscious application of the evaluation criteria and the deliberate movement toward systems that justify the trust investment they require. Teams that use these pillars as consistent measurement tools achieve better outcomes even when individual implementations are imperfect.

Consider what this looks like in practice. While some build AI that sits on top of existing operating systems, **Venho.AI** represents something fundamentally different: an operating system built from the ground up around human intent, with every design decision reflecting conscious choice between enhancement and extraction.

A Model of Conscious Choice

Conscious choice isn't just architectural—it's philosophical. In January 2024, during a trip to Africa, Venho.AI founder Antti Saarnio had an epiphany: "Everybody should have their own AI. The future can't be that it's somebody's in somebody's cloud. The data actually should be yours as well, sitting in your computer or device."

This insight reflects Saarnio's distinction between "human-centric AI" and "resource-centric AI." As he explains: "There's a difference between the U.S. approach, which is now the big tech approach, basically like an oil industry approach where people are resources, and the European approach, traditionally, where you need to empower people so they become contributors."

That philosophical choice flows into every design decision. Working with cognitive psychology professor Göte Nyman, the team developed "active memory"—user-defined memory items in which you explicitly direct what your AI should attend to, when, and for how long. Instead of systems that learn from behavior to optimize engagement, you control what the AI learns about your intent.

This represents conscious choice at every level. Saarnio could have built automated systems that predict user needs, but deliberately chose human-in-the-loop design: "I wanted to keep user control."

Instead of cloud-based processing for data collection, all processing occurs locally. Rather than relying on advertising revenue, users pay directly. As Saarnio puts it: "You're the proprietor, not the product."

The **Jolla Mind2** device, powered by Venho, demonstrates this philosophy in practice: a personal AI companion that processes locally while connecting to the user's broader digital ecosystem. Users interact via natural-language commands rather than app switching, providing seamless access to AI capabilities without surrendering control.

"We think mission first," Saarnio explains. This mission-first approach shows what becomes possible when teams consistently apply evaluation criteria that prioritize human flourishing over corporate convenience.

Throughout their development process, Venho's team has navigated the same tensions that every AI team faces—and their solutions reveal what each of the Six Pillars looks like in practice when applied consistently under real-world dynamics.

But principles alone don't create sustainable outcomes when market forces bear down on teams. Progress is not a guarantee. Teams need guideposts for maintaining those principles under business pressures. They need evaluation criteria that make conscious choice easier than standard optimization approaches.

Guideposts, Not Guarantees

The Six Pillars of Intent serve a specific purpose: they provide shared language for evaluating how AI implementations affect human capability, agency, and flourishing.

These aren't technical specifications or product roadmaps. The pillars are evaluation criteria—reference points for navigating decisions made by product teams, business leaders, and individual users, collectively shaping how AI is developed and deployed.

Every day, countless decisions determine whether technology will understand and serve human intent or exploit human psychology for engagement and profit. Most of these decisions are made without anyone explicitly considering the basic relationship to technology.

Engineers might optimize for performance. Designers could solely focus on user experience. Executives may pursue market opportunities. But those choices, taken together, create the patterns society lives with.

The pillars create a shared vocabulary for recognizing these patterns as they unfold:

Human Connection Enhancement asks: *Does this reduce digital noise, creating space for meaningful human interaction? When AI handles routine coordination, does it make more quality time for meaningful human relationships?*

Trust-Centered Design asks: *Does this preserve user agency over data and decisions? Can users understand how this system operates and modify its behavior when necessary?*

Seamless Integration asks: *Does this reduce the mental burden of managing multiple systems while preserving user control over workflows? When AI coordinates across digital tools, do users retain authority over how those tools interact?*

Anticipatory AI asks: *When systems predict users' needs, do they strengthen users' ability to achieve their goals? When systems take action before users request it, do users remain architects of their own experience?*

Mobile as AI Gateway asks: *Does this make advanced AI features accessible anywhere without compromising privacy? Can users access contextual intelligence through natural interaction?*

Environmentally Responsible Innovation asks: *Does this system use computing power efficiently? Were the environmental costs weighed against genuine human benefit, and is there transparency to that?*

These questions provide a practical approach to maintaining human-serving principles under societal or market pressures. The urgency isn't whether to apply them, but how quickly. Every passing quarter narrows the window for conscious choice.

The Window for Conscious Choice

The rapid deployment of AI has surpassed our ability to evaluate whether these systems align with human intent or primarily serve corporate interests. Every quarter, more organizations integrate AI into workplace tools, customer interactions, and personal devices, often without clear guidelines to differentiate between enhancement and exploitation. The infrastructure choices being made today—regarding the location of intelligence, the flow of data, and which business models receive funding—will have lasting impacts for decades and will shape the technological landscape that future generations will inherit.

But this acceleration creates both risk and unprecedented opportunity. For the first time, the tools for conscious choice exist simultaneously: technical capabilities that enable privacy-preserving AI, growing market recognition that extraction models are unsustainable, and frameworks for systematic evaluation. Teams like Antti Saarnio's at Venho.AI demonstrate that privacy-protecting systems, context-aware assistance, transparent recommendations, human-in-the-loop decision making, and responsible resource use aren't future concepts—they're engineering choices being implemented today.

The Six Pillars serve as essential criteria for measuring whether AI implementations enhance human capability while preserving individual agency and honoring the decades of trust that users have invested. Just as the marketing technology capabilities continued to thrive after my departure from the digital agency—because the team embraced the approach rather than merely executing it—these pillars are effective when teams internalize the evaluation method rather than relying on vague assumptions. They ensure that solutions address genuine problems rather than creating complex solutions for problems that may not exist.

The window for conscious choice narrows because the feeling of inevitability becomes normalized with each new deployment. When organizations choose convenience over agency, engagement metrics over authentic connection, and algorithmic efficiency over human intent, the path of least resistance moves further from human-centered outcomes. Each quarter of deferred conscious choice makes human-centered alternatives more expensive to implement and harder to justify against entrenched competitive dynamics.

Organizations that master intent-driven evaluation now don't just build better products—they position themselves for an economy where conscious choice becomes the competitive differentiator. Those who wait until patterns of extraction dominate their markets will find themselves competing against industry momentum rather than redirecting it.

Each pillar reveals different dimensions of conscious choice in AI development, exposing how technology can either create space for human flourishing or undermine it. Understanding how these principles translate into specific design decisions—and how strategic leaders, product teams, and policymakers can apply them today—requires examining each pillar as both a guidepost and a practical methodology. The framework becomes powerful when organizations can recognize these patterns in their daily decisions and advocate for better human-centered outcomes.

Chapter 5 -

Pillar 1: Human Connection Enhancement

When Presence Disappeared

The fortieth anniversary of the movie *The Breakfast Club* was recently celebrated, with the cast getting together at the Chicago Comic & Entertainment Expo (C2E2). Molly Ringwald, Michael Anthony Hall, Judd Nelson, Emilio Estevez, and Ally Sheedy were doing a panel where an audience member asked, "What is something, physical or metaphorical, you would want to bring back from the '80s for kids today?"

Ringwald immediately answered, "No cell phones." She got unanimous support from her castmates, with Estevez adding, "We spent our time looking up rather than down, and we saw more of the world as a result of that."

The actors helped shape how a whole generation looks at authentic relationships, yet their almost instinctive response wasn't about music, fashion, or cultural trends. It was about removing the technology that now commands hours of daily attention.

Their longing wasn't analog nostalgia—it was about getting back to having people's full attention while talking to them, being fully present in the moment, and making connections without needing a digital middleman. They mourned moments when being together meant actually being together, without notifications interrupting dinner or screens competing for engagement.

The Breakfast Club was fundamentally about breaking down barriers and seeing each other for who we are. Five teenagers from different social cliques discovered they had more in common than they thought. The movie's enduring power stems from recognition that humans all struggle with the same basic need to be understood and accepted for who they truly are.

The cast's responses suggest concern about inhabiting a distracted digital environment that makes it harder, rather than easier, to see people for who they are. Research supports their intuition: smartphone users check their devices a global average of 58 times per day, though this varies significantly by age group and region.

This level of device distraction would have been inconceivable in the 1980s but has become normalized today. The technology designed to bring us closer together has become a primary barrier to authentic human presence. I believe the problem isn't that smartphones exist—it's how most current systems prioritize engagement over enhancing relationships, and capture attention over offering genuine value.

The attention economy treats human focus as a resource to be extracted rather than protected. Social media platforms employ psychological tactics to elicit dopamine responses, keeping users scrolling even when the experience often leaves them feeling empty.

Notification systems interrupt thoughts hundreds of times daily, training expectations for constant stimulation, and making sustained attention increasingly difficult.

Studies show that when attention is constantly fractured, people lose the capacity for the deep presence that authentic connection requires. The mental switching between digital demands and human relationships isn't just individually exhausting—it erodes our ability to see and understand one another as human beings.

The key challenge lies in distinguishing between AI systems that foster genuine human relationships and those that exploit psychological vulnerabilities to boost engagement metrics. This pivotal moment in AI development presents a unique opportunity to address this issue through conscious design choices that prioritize human intentions over merely extracting attention.

Defining Human Connection Enhancement

Human Connection Enhancement serves as an evaluation criterion for assessing how AI systems affect authentic human relationships. This pillar asks a fundamental question: *Does this reduce digital noise to create space for meaningful human interaction while protecting the sustained attention that authentic connection requires?*

The pillar measures progress along three critical dimensions that teams can apply immediately to their AI implementations:

Attention Quality Over Attention Quantity. Current technology fragments focus through constant interruptions designed to maximize user engagement. This dimension assesses whether AI systems protect mental energy by intelligently filtering noise to create uninterrupted time for meaningful relationships and focused work.

User Agency Over AI Learning. Technology intended to connect people is most effective when individuals have control over how AI learns about their relationships and social patterns, versus systems that automatically collect behavioral data to enhance engagement. This dimension measures whether individuals can control what their AI remembers about their social connections and modify the AI's behavior as needed.

Bridging Differences Not Exploiting Them. Technology has the potential to either enhance our understanding of one another—by accounting for diverse communication styles, cultural backgrounds, and individual differences—or to deepen divisions through automated manipulation. This dimension focuses on whether AI systems promote understanding of diverse human experiences or reinforce isolation within ideological echo chambers.

The evaluation framework acknowledges that human connections exist on a spectrum of intimacy and purpose. Family relationships require different types of support compared to professional collaborations, which, in turn, differ from community engagement or casual social interactions. Effective AI systems recognize these varying situations and do not apply the same approach uniformly across all human relationships.

Strategic leaders can apply this pillar by asking whether proposed AI implementations will help their teams collaborate more effectively or create additional digital overhead. **Product teams** can evaluate whether features enhance user relationships or exploit psychological vulnerabilities. **Policymakers** can assess whether regulations encourage technologies that bring people together or enable platforms to profit from keeping them divided.

The pillar also addresses the misleading notion that we must choose between technological advancement and human authenticity. Enhanced connectivity through AI does not equate to automating human interaction; it creates conditions in which genuine human connections can thrive without digital interference or manipulation. Unlike Character.AI's approach, which exploited synthetic intimacy with tragic consequences for vulnerable users, true connection enhancement fosters authentic human bonds.

When AI manages user-guided routine coordination and information filtering, individuals gain the capacity for sustained attention that is essential for meaningful relationships.

This evaluation criterion becomes essential as AI capabilities scale globally through mobile devices. The same technologies that could eliminate barriers between people could also perfect the attention extraction that already fragments human presence. Human Connection Enhancement provides measurement criteria to ensure AI development serves authentic relationship-building rather than sophisticated manipulation.

The challenge organizations face is not whether to build AI capabilities, but whether those capabilities will enhance human connection or replace it with artificial digital interactions designed to maximize corporate engagement metrics.

Conscious Connection

Venho.AI demonstrates Human Connection Enhancement through design choices that prioritize user agency over optimizing for user engagement. When founder Antti Saarnio decided to build an AI operating system from the ground up around human intent, every design choice reflected a conscious commitment to enhancing rather than extracting human attention.

The contrast with comparable AI systems becomes immediately apparent in Venho's approach to memory and learning. While many platforms automatically collect behavioral data to optimize engagement, Venho.AI created "active memory"—user-defined memory items that allow individuals to explicitly direct what their AI should learn, when, and for how long. Users control what the system remembers about their intentions, rather than having algorithms infer patterns from surveillance.

"I wanted to keep user control," Saarnio explains. This philosophical choice cascades through every system component. Instead of cloud processing that enables data collection, all processing occurs locally on user devices. Instead of advertising revenue that requires attention capture, users pay directly for capabilities. Rather than maximizing time-on-device, Venho optimizes for user satisfaction and task completion.

This approach directly addresses attention fragmentation, which undermines human connection. Rather than competing for focus through notifications and engagement tactics, the system becomes virtually invisible during human interactions. The Jolla Mind2 device, powered by Venho, exemplifies this concept: Users can interact via semantic commands rather than constantly switching between applications. This reduces cognitive switching while allowing them to retain complete control over their digital environment.

The implications of human connection extend beyond the individual user experience. When Saarnio describes the choice between "resource-centric AI" and "human-centric AI," he identifies a fundamental design philosophy: Whether technology should extract value from human attention or enhance human capability.

Resource-centric approaches treat people as sources of data and engagement to be optimized. Human-centric design treats technology as a tool that should fade into the background, creating space for authentic presence and meaningful relationships.

Venho.AI's "mission-first" approach illustrates how teams can maintain Human Connection Enhancement principles even under market pressure to optimize for traditional engagement metrics. When business models align with user interests through direct payment rather than surveillance capitalism, the entire development trajectory shifts toward serving genuine human needs rather than exploiting psychological vulnerabilities.

The technical architecture reflects these values. Local processing means users retain control over their data and computing power. Semantic interaction reduces mental switching, which fragments attention during human conversations. User-defined learning parameters prevent the system from developing behavioral predictions that could be used for manipulation.

This represents a conscious choice at every design level. Saarnio could have built systems that automatically predict user needs, but deliberately chose a "human-in-the-loop" design. He could have implemented cloud processing to enhance performance, but prioritized user control over corporate convenience. He could have pursued advertising revenue models, but chose direct payment to align business incentives with user interests.

The design choices underlying Venho demonstrate how Human Connection Enhancement serves the trinity of forces that define the Mobile Era of Intent. **Agency** is maintained because users control their attention allocation, interaction parameters, and memory configurations without any automated manipulation. **Trust** develops through transparent local processing that doesn't require users to surrender their behavioral data to systems designed for extracting engagement. **Intent** serves as the guiding principle, allowing technology to fade into the background during human interactions, enhancing presence and connection rather than competing for attention through notifications that disrupt focus.

The result demonstrates that Human Connection Enhancement isn't just theoretical—it's achievable through deliberate design decisions that teams can implement today. When AI systems understand and serve human intentions rather than merely extract human attention, they create conditions in which authentic connection can flourish without digital interference.

Human Connection Enhancement in Practice

The Human Connection Enhancement criterion becomes clearer by examining other real-world implementations in which design choices affect our capacity for authentic relationships. The following examples demonstrate how design decisions can produce distinctly different outcomes in human attention and connection quality.

Wordly AI's real-time translation demonstrates how AI can eliminate barriers to authentic democratic participation. The system enables multilingual communication across 60+ languages during city council meetings and community forums, allowing non-English speakers to fully engage in civic discussions. Municipalities using Wordly report significant increases in participation from immigrant populations and in compliance with accessibility requirements. The technology dissolves language barriers without requiring users to compromise their authentic voice or cultural expression.

Aidoc's empathy-enhanced medical imaging illustrates AI that amplifies rather than replaces human judgment. The system identifies not only medical anomalies but also indicators of patient anxiety extracted from communication patterns. When distress is detected, it prompts clinicians to respond more empathetically during difficult conversations. A 2024 study found a 30% improvement in patient trust ratings among facilities that used these modules. The AI creates space for a deeper human connection by acknowledging user stress while preserving the clinician's role in providing care and compassion.

Sephora's emotion-aware customer service shows how AI can recognize human emotional states without manipulating them. The system detects frustration or confusion in real-time chat interactions and adjusts tone, pacing, and content accordingly. When emotional distress is identified, conversations get escalated to human representatives who can provide appropriate support. Customer satisfaction surveys report a 22% increase in brand satisfaction and a reduction in complaint escalation. The AI serves human needs by recognizing when human intervention becomes necessary.

These implementations demonstrate how conscious design can create sustainable competitive advantages through genuine user empowerment. Protecting focus, preserving agency, and facilitating authentic connection aren't compromises—they're strategic choices that align business success with human flourishing.

However, the same AI capabilities that could restore authentic human connection can just as easily perfect attention extraction when different design priorities prevail.

Tinder's automated matchmaking reveals how AI can exploit human psychology to drive corporate metrics rather than serve genuine compatibility. The system optimizes for "return swipe probability" and time-on-app rather than relationship quality, using reinforcement learning to keep users engaged in "swipe loops." Research characterizes this as "algorithmic colonization of love," in which users report emotional fatigue and inauthentic behavior as algorithms train them to optimize for engagement rather than authentic connection.

HireVue and **Workday**, AI recruitment screening companies, demonstrate how automation can eliminate human judgment from relationship-building processes. These systems optimize for resume keywords and behavioral signals from video interviews, often filtering out qualified candidates with non-traditional backgrounds. A 2024 audit revealed excessive disqualification of minority candidates, while organizations report poor cultural fit when human intuition is removed from hiring decisions. The AI optimizes for automated performance rather than the human potential that meaningful work relationships require.

AI-generated content "slop" illustrates the widespread decline of authentic human expression. Research by SEO firm Graphite shows that AI-generated content now comprises over 50% of internet articles, much of it low-quality, repetitive, or misleading, designed to exploit engagement algorithms. This flood of synthetic content makes it increasingly difficult to distinguish genuine human communication from automated noise, fragmenting attention and undermining trust in digital discourse. The proliferation crowds out authentic voices in favor of content optimized for search rankings and advertising revenue.

These contrasting implementations highlight practical criteria that teams can apply right away. Human Connection Enhancement addresses several key questions: Do AI systems help preserve cognitive resources for meaningful interactions, or do they overwhelm users with information? Are users in control of what AI learns about their relationships, or do systems automatically gather behavioral data for the purpose of capturing attention? And, does technology facilitate understanding among individuals from different backgrounds, or does it exploit societal divisions for profit?

The most important insight comes from analyzing the business models behind these choices. Systems that receive funding through direct payments or genuine value creation typically promote human connection, while those that rely on attention capture or data extraction often exploit it. When AI teams prioritize user satisfaction over engagement metrics, technology can enhance genuine relationships rather than diminish them.

But even well-designed features can be undermined by the larger technological systems they operate within. Human Connection Enhancement requires more than conscious design choices by separate teams—it demands evaluation frameworks that can be applied consistently across the complex ecosystems that shape today's digital experiences.

However, Human Connection Enhancement alone cannot address the full complexity of AI systems that serve human intent. Protecting attention from extraction requires trust that technology serves users' interests rather than surveils them. Reducing cognitive overhead requires the Seamless Integration of various platform features rather than creating new silos.

Understanding human intentions necessitates anticipatory capabilities that preserve rather than replace human agency. These connections reveal why evaluation frameworks require multiple measurement criteria to operate in concert rather than focusing on a single pillar.

Evaluating Human Connection-Enhancing AI

Human Connection Enhancement becomes actionable when organizations can apply consistent measurement criteria to their AI implementations. The framework provides practical questions that strategic leaders, product teams, and policymakers can use to distinguish between systems that foster authentic relationships and those that manipulate user behavior.

For Strategic Leaders

Strategic decision makers can evaluate AI investments by examining whether proposed systems will amplify their organization's capacity for meaningful stakeholder relationships or optimize for metrics that undermine long-term trust.

Does this AI implementation create more time and mental space for our teams to engage in high-value human interaction? Systems that manage routine coordination, filter out information noise, and minimize repetitive tasks help free up mental energy for creative collaboration and relationship-building. However, implementations that introduce new dashboards to monitor, additional alerts to manage, or complex interfaces to learn can increase cognitive burden rather than alleviate it.

Will this technology help our customers, partners, or employees better understand one another despite differences in communication styles, cultural backgrounds, or levels of expertise? AI that connects diverse aspects of humanity enhances relationships. Systems that reduce diversity to algorithmic bias or manipulate differences for engagement metrics distort connections.

How does the business model underlying this AI align with our stakeholder relationships? Technologies that are funded by direct user payments typically foster human connection, while those that rely on capturing attention or extracting behavioral data often exploit it. Direct payment models align providers' interests with user satisfaction, whereas advertising-funded systems primarily focus on maximizing clicks rather than on relationship quality.

For Product Teams

Product developers can apply Human Connection Enhancement criteria during the design and testing phases to ensure that AI features serve authentic user needs rather than merely corporate engagement metrics.

Does our AI help users accomplish their goals more efficiently so they can focus on what matters most to them personally? Effective systems reduce the time and attention required for routine tasks, creating space for relationships, creativity, and meaningful work. Features that increase session duration or daily active usage might optimize for engagement at the expense of user satisfaction.

Can users easily understand what our AI is doing and maintain meaningful control over its behavior? Transparency enables trust, while user control preserves agency. Systems that operate as black boxes or manipulate user behavior through opaque recommendation algorithms undermine the foundation for authentic relationships with technology.

How does our AI perform across diverse user populations, communication styles, and cultural contexts? Inclusive design ensures that technology bridges existing divides rather than amplifying them. AI systems that accurately represent dominant demographics but neglect marginalized communities create barriers to equitable participation in digital spaces.

For Policymakers

Policy frameworks can encourage Human Connection Enhancement by creating incentives for AI development that prioritize human flourishing over widespread exploitation.

Do current regulations require transparency about how AI systems make decisions that affect human relationships, such as hiring, lending, or content recommendations? Policy frameworks that mandate AI accountability enable conscious choice about when and how to engage with AI-mediated systems.

Are there mechanisms for individuals to seek recourse when AI systems make errors that damage relationships or opportunities? Effective oversight ensures that AI deployment serves justice rather than simply efficiency, particularly when automated decisions affect employment, housing, or access to services.

How do privacy and data protection laws align AI business models with user interests rather than surveillance capitalism? Regulatory approaches that require explicit consent for behavioral data collection and provide users control over their digital profiles encourage AI development that serves rather than exploits human psychology.

Measurement Criteria

Teams can measure progress through concrete metrics that reveal whether their design choices serve authentic human connection or merely extract attention.

Attention and Information Quality: *Do users experience more uninterrupted time for meaningful relationships and focused work, or do users spend more time managing digital experiences despite AI assistance?* Enhancing genuine connection protects cognitive energy for deep engagement rather than creating new forms of digital distraction.

User Agency Over AI Learning: *How much meaningful control do users maintain over what AI systems learn about their relationships and social patterns, and how does this accommodate different privacy preferences?* Connection-enhancing systems enable users to direct AI learning about their social interactions rather than automatically collecting behavioral data for engagement metrics.

Bridging Differences Rather Than Exploiting Them: *What degree of understanding do AI systems facilitate between people with different communication styles, backgrounds, and perspectives?* Human Connection Enhancement helps people understand one another across differences rather than exploiting divisions to harvest attention.

When organizations make Human Connection Enhancement a standard evaluation criterion—applied consistently from initial system design through ongoing feature development—they support more purposeful technology decisions. Teams can then distinguish between technology that manipulates users for corporate benefit and technology that succeeds by genuinely helping people.

The evaluation criteria provide shared language for making conscious choices about how AI develops, but implementation requires more than individual team decisions. Human Connection Enhancement depends on technology systems that support authentic relationships rather than fragmenting them.

Technology That Brings Us Together

When Molly Ringwald instinctively answered "no cell phones" as what she'd bring back from the '80s for today's kids, she identified the critical choice society faces with AI. The technology intended to connect people has become a primary barrier to authentic presence. AI development has the real potential to help reclaim the undivided attention that meaningful relationships require, or to perfect the attention extraction that already fragments human connections at an unprecedented scale.

The difference lies entirely in how AI systems are designed and deployed. When teams consistently apply the Human Connection Enhancement evaluation criteria, individual benefits add up, strengthening society as a whole. When people gain access to AI that protects rather than exploits their attention, they become more present with family, more focused at work, and more engaged in their communities.

Consider the cascading effects: AI that reduces coordination overhead enables parents to spend quality time with their children rather than managing digital logistics. Workplace AI that eliminates administrative friction enables teams to find creative solutions to complex problems rather than battle with information overload. Civic AI that breaks down language barriers enhances democratic participation.

But the same AI capabilities that could restore authentic connection also threaten to eliminate it. Systems designed around engagement metrics could perfect the behavioral manipulation pioneered by social media. AI that understands human psychology with unprecedented precision could exploit rather than serve our deepest needs for connection and meaning.

The stakes mirror what happened in that Saturday detention. Those teenagers discovered something profound: beneath surface differences and social performances, humans share fundamental needs for understanding, acceptance, and genuine connection. AI development in the Mobile Era of Intent can enable similar discoveries on a global scale—technology that helps us see each other clearly across differences, that makes space for authentic connection rather than engineered engagement.

This vision requires a conscious choice of evaluation criteria applied consistently throughout AI development. Human Connection Enhancement provides one lens for making these choices consciously, but it is most effective when combined with the other pillars of intent-driven AI. Authentic relationships require not just protection from attention extraction, but also Trust-Centered Design experiences that serve our interests, Seamless Integration that reduces rather than increases mental burden, and Anticipatory AI capabilities that proactively address intent while preserving human agency.

Chapter 6 -

Pillar 2: Trust-Centered Design

When Trust Was Lost

A simple oil change, a dropped coolant cap, and a lie that was never acknowledged. That was all it took to end my relationship with a local auto care business I had trusted for years.

I was sitting in the waiting area, half-watching through the service bay window as a technician worked on my car. The routine was familiar: drain the oil, replace the filter, top off the fluids. I'd been bringing my vehicles here for five years. Friendly staff, reliable service, and continued trust that they'd take care of my needs.

Then I watched the technician accidentally drop my car's coolant cap into the engine compartment. I saw him look around, clearly flustered, before walking away toward the parts area. When he returned, he installed what was obviously a new cap, not the one he had dropped.

Fifteen minutes later, he brought me the final tallied bill with an additional $25 charge for a "missing coolant cap."

I stared at the line item, then looked up at him. "Um, you dropped my cap down into the engine. Why am I being charged for a new one?"

The moment stretched. He could have laughed it off, apologized for the mix-up, and removed the charge. Instead, he doubled down. He told me the cap was already missing when they started, and it had to be replaced for safety.

I pointed to the engine bay. "It's still in there."

Without a word, he went back, fished out my original cap from the engine compartment, and removed the new cap from the estimate. He never apologized.

I paid the bill. I never went back.

Here's why this matters. It wasn't about the extra $25 on the bill. The oil change was done correctly. My car ran fine. But something more significant than a plastic cap had been lost. Trust was broken by a choice to deceive rather than admit a simple mistake.

That moment also revealed that trust relationships are remarkably fragile and more complex than mere betrayal or the discontinuation of a service. I could have returned despite the deception—maybe they were the only convenient option, or finding a reliable mechanic was too complicated, or the overall service quality outweighed one dishonest moment. Trust isn't binary. It's a nuanced calculation involving available alternatives, proximity, pricing, and scheduling convenience, and I decided those factors didn't justify staying.

Now imagine that same trust dynamic, but instead of a local mechanic, it's an AI system that knows daily routines, personal preferences, financial decisions, and the family's schedule. Instead of a $25 overcharge, it's subtle nudges toward purchases you don't need, recommendations that serve corporate interests rather than your own, or decisions made on your behalf that you never explicitly authorized.

AI design teams face similar choices about trust dynamics at the system level that could affect millions of users simultaneously. When systems collect behavioral data, will that information serve user goals or enable surveillance? When algorithms make recommendations, will they optimize for user satisfaction or engagement metrics? When AI processes personal data, will users retain meaningful control, or become captive to opaque systems?

These aren't abstract philosophical questions. Business decisions are being made in product meetings, reflected in code, and embedded in infrastructure that shapes daily digital experiences. The difference lies in scale and consequence: the mechanic's choice affected one customer relationship, while AI design decisions can affect millions of users, leaving little practical recourse when trust is violated.

The trust people have invested in digital systems over decades forms the foundation for AI adoption today. Whether willingly or reluctantly, users adapted to new interfaces, shared personal information for convenient services, and accepted technological complexity in exchange for promised benefits.

That accumulated trust can either fund genuine progress toward systems that understand and serve human intent, or it can be extracted through surveillance, manipulation, and automated exploitation.

Technology features don't predetermine the choice between these paths. It's determined by conscious decisions about how AI systems are designed, deployed, and governed.

Defining Trust-Centered Design

Trust-Centered Design serves as a criterion for distinguishing AI systems that honor user trust investments from those that exploit it for corporate advantage. This pillar asks a fundamental question: *Does this technology deserve the trust it requires to function effectively, and does it give users meaningful agency over that trust?*

The pillar measures progress along three critical dimensions that teams can apply to their AI implementations:

Transparency That Enables Control. Users can understand how AI systems make decisions that affect their lives, and they maintain meaningful control over modifying system behavior when it doesn't align with their intentions. This requires genuine user agency—the practical capacity to understand, contest, and redirect AI decisions rather than merely being informed afterward.

Incentive Alignment Through Design. The business model and technical architecture create structural reasons for the AI system to serve user interests rather than exploit them. When users pay directly for capabilities, when processing occurs on their devices, when data serves their goals rather than corporate priorities—these architectural design decisions make Trust-Centered Design systemic rather than aspirational.

Recourse and Reversibility. Users can seek meaningful remedies when AI systems make errors, and they can exit relationships without losing their data, digital identity, or access to essential services. Trust becomes measurable through user autonomy—the practical ability to understand, control, and, if necessary, exit AI-mediated relationships without leaving digital residue as training data.

The evaluation framework recognizes that trust requirements vary across different contexts and relationships. Enterprise AI serving business workflows needs transparent decision records and clear accountability structures. Consumer AI managing personal information requires intuitive user controls and data freedom. Healthcare AI systems that make diagnostic recommendations require explainable decision making and clinical oversight. Each context has distinct trust requirements, but all require systems that users can understand, control, and exit when necessary.

Strategic leaders can apply this pillar by assessing whether proposed AI investments enhance or diminish user agency over time. **Product teams** can evaluate whether their systems empower users to make informed decisions or manipulate their behavior through opacity. **Policymakers** can assess whether regulations promote systems that empower users or enable sophisticated forms of digital manipulation.

The pillar addresses the misconception that trust and innovation exist in tension. Trust-Centered Design systems often demonstrate superior technical performance because they must solve genuine user problems rather than optimizing for engagement metrics that may conflict with user satisfaction. When AI teams optimize for user empowerment rather than user capture, they create sustainable competitive advantages based on earned loyalty rather than making it difficult for users to leave.

This evaluation framework becomes essential as AI capabilities expand through mobile devices, which users rely on for their most personal functions. While these technologies have the potential to empower users in their digital relationships, they also risk perfecting surveillance and manipulation on an unprecedented scale. Trust-Centered Design offers measurement criteria to ensure that AI development enhances, rather than undermines, the trust relationships that drive innovation.

Conscious Trust Architecture

Venho.AI prioritizes human augmentation over automated engagement, and this commitment is reflected in design decisions that build rather than exploit user trust. When Antti Saarnio decided to create an AI operating system, he faced fundamental questions about control dynamics: Who controls the AI? Who sees the data? Who makes the decisions?

"Human in control at every significant phase of AI actions," Saarnio explains, describing Venho's core trust principle. While AI handles email processing, calendar management, and workflow automation in the background, users retain approval authority over all final actions. This isn't a user experience design choice—it's a trust architecture that prevents what Saarnio calls the nightmare scenario: "AI running wild."

The distinction is important because, while AI systems may strive for seamless automation to reduce friction, doing so can diminish meaningful human involvement. Venho intentionally introduces user agency at key decision points, requiring users to approve AI recommendations rather than imposing automated control. When the system suggests prioritizing certain emails or scheduling meetings, users can see the reasoning behind these suggestions and maintain the power to override them.

This control extends to memory management, where Venho takes what cognitive scientist Göte Nyman calls a "personal perspective to data." Rather than using database logic, the system learns how users think about their work and priorities. Users control this learning, defining which patterns and connections matter.

"Building and maintaining a trusted personal database for AI use," as Nyman describes it, means users understand how their data is categorized, accessed, and applied. The AI doesn't develop mysterious insights about user behavior—it develops transparent models that users can inspect and modify.

Saarnio's choice of local processing over cloud services reflects a trust-centered approach to data sovereignty. By designing Venho around an efficient 3B parameter model, local processing becomes both more cost-effective than cloud usage fees and more private than remote data processing. Users control their information without trusting corporate privacy policies or government data requests. The design itself serves as the privacy guarantee, rather than relying on the promises of technology companies.

The subscription model reinforces these trust dynamics by aligning business incentives with user interests. Rather than generating revenue through advertising or data monetization, Venho succeeds when users find the system valuable enough to pay for continued access through a monthly subscription. This creates aligned incentives, meaning the company's financial health comes from genuine user satisfaction rather than engagement metrics that may conflict with user well-being.

Venho's architectural choices reveal how Trust-Centered Design serves the trinity of forces that define the Mobile Era of Intent. **Agency** is retained because users maintain control over AI decision-making authority, data processing locations, and memory configurations, without automated oversight. **Trust** develops through transparent local processing and inspectable models that don't require surrendering control to opaque cloud systems or deciphering convoluted privacy policies. **Intent** serves as the organizing principle that enables users to maintain meaningful involvement in AI actions while benefiting from automation—the system serves their goals through collaboration rather than making decisions on their behalf.

These technical choices demonstrate that trust-centered AI requires conscious design decisions that often conflict with standard optimization targets. Saarnio could have built a more autonomous system, processed data in the cloud, or monetized user behavioral data, but chose transparency and user control instead. The result validates that Trust-Centered Design is achievable through deliberate technical design and business model choices.

Trust-Centered Design in Practice

Trust-centered and exploitation-focused AI becomes distinguishable when examining how different organizations approach similar AI implementations. The same AI features that empower users can just as easily enable widespread exploitation, depending on the design choices made during system design and development.

Recent implementations reveal striking contrasts between systems designed to honor user agency and those optimized for corporate control. When organizations face decisions about data handling, business models, and user interfaces, their choices either strengthen or erode the trust investments on which technological progress depends.

Hugging Face Model Repository demonstrates community governance in AI development, prioritizing transparency over corporate control. The platform provides detailed model cards that reveal the training data sources, limitations, and bias evaluations, allowing users to verify them independently. While this approach requires navigating complex trade-offs between open standards and security, it enables researchers to inspect, modify, and deploy models without relying on providers' decisions about access or pricing. Users retain control over their AI capabilities rather than relying on third-party services.

Local-First AI tools such as Ollama and Jan.ai demonstrate trust through user control and technical privacy. These systems run AI models entirely on users' devices, making external data collection technically impossible, rather than merely prohibited. Users retain full control over their conversations, documents, and model selection without relying on external services. The open-source nature enables security audits and community verification while eliminating subscription fees and usage limits that could create future service dependencies.

Proton Scribe AI Assistant shows how enterprise AI can prioritize privacy by design. Built by the same team behind Proton's encrypted email and storage services, Scribe processes documents and generates content without sending data to external servers. The Swiss-based company operates under strict privacy laws and maintains a business model funded by user subscriptions rather than data monetization, aligning corporate incentives with user protection.

These approaches demonstrate how Trust-Centered Design can create sustainable competitive advantages. Community governance, local processing, and privacy by design aren't compromises—they're strategic choices that align business success with user empowerment.

However, the same technological capabilities can just as easily enable systematic exploitation when different business model decisions prevail.

Adobe Firefly's Training Deception violated trust through marketing misrepresentation. Despite promoting Firefly as "commercially safe" and training it solely on licensed Adobe Stock images, Bloomberg reported that 5% of the training data came from AI-generated content produced by competitors such as Midjourney. Adobe never disclosed this publicly, even as it claimed ethical superiority over "openly scraped" competitors. Enterprise customers who paid premium prices for legally defensible AI content discovered they faced potential copyright liability despite Adobe's promises. Internal employees raised ethical concerns about using competitor content, yet the company continued external marketing, emphasizing ethical data training.

Harvey AI Legal Hallucinations demonstrated how AI systems designed to appear authoritative can create professional liability without adequate safeguards for verification. The AI legal research service provided fabricated case citations to lawyers preparing court filings, resulting in sanctions and professional embarrassment when judges discovered the non-existent precedents. While the lawyers failed to perform the basic verification required by legal training, the system's interface design suggested authoritative reliability, encouraging overreliance without fact-checking. The design itself created conditions for professional liability.

Clearview AI Facial Recognition exemplifies a trust violation by eliminating consent and creating surveillance infrastructure. The company scraped billions of photos from social media platforms without user consent, then sold facial recognition services to law enforcement without public oversight or accountability. Users had no knowledge that their images were being used, no control over the technology's deployment, and no recourse when the system enabled mass surveillance capabilities that fundamentally altered the balance between privacy and state power.

These failures weren't accidental—they followed from design and business model choices that prioritized corporate benefit over user protection. Adobe could have excluded AI-generated content from training data. Harvey AI could have implemented confidence scoring and uncertainty alerts when citing case precedent. Clearview AI could have sought consent and operated with transparency. The technology existed to make different choices. The incentives pointed toward extraction instead of honoring user trust investments.

The pattern reveals that trust in AI systems depends less on what providers promise than on design decisions that either enable or constrain harmful use. When systems are designed to maximize corporate control, user dependency, and data extraction, trust violations become the norm rather than the exception.

The Trust Exchange Framework

Trust in AI systems operates through exchanges between three critical parties: the people who use these systems, the providers who build them, and the policies that govern their interaction. Each invests something of value and expects something in return. When these exchanges honor the investments made, trust accumulates, enabling more sophisticated AI capabilities. When they violate those investments, trust erodes and creates barriers to beneficial innovation.

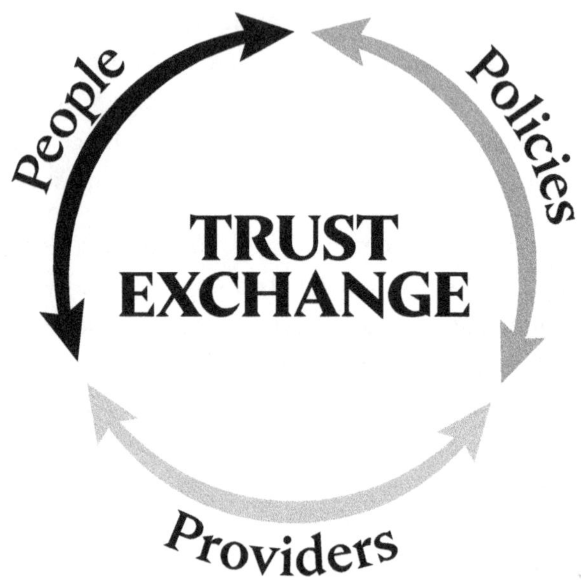

Figure 6.1: The Trust Exchange - Critical interconnected parties of People, Policies, and Providers

This isn't abstract theory—it's the practical reality that determines whether AI systems thrive or fail. I developed this framework through observing a consistent pattern: The Trust Exchange is never static. Any of the three parties can give trust willingly or reluctantly, or withdraw it entirely, and those shifts create cascading effects throughout the engagement.

When people reluctantly accept AI systems because alternatives aren't available, providers may mistake compliance for satisfaction. When providers reluctantly comply with policies while seeking workarounds, genuine trust-building grinds to a halt. When policies reluctantly accommodate rapid AI deployment without adequate safeguards, people's trust in regulatory effectiveness erodes.

The quality of trust investment—willing versus reluctant—determines whether these exchanges create sustainable progress or fragile dependencies that collapse under pressure. The framework explains why some AI implementations can build lasting relationships, while others could provoke backlash and abandonment.

The three parties invest different trust currencies:

People invest in the belief that AI systems will serve their actual goals rather than exploit their engagement for the provider's benefit. They provide adoption and behavioral feedback that validates whether systems work, along with the social acceptance that makes widespread deployment viable.

Providers invest capital, technical expertise, and reputation in exchange for people's trust and market acceptance. They make conscious choices regarding privacy, transparency, and business model alignment that determine whether their systems retain user confidence or face user withdrawal.

Policies invest regulatory authority and enforcement resources in exchange for technological innovation that serves people's interests rather than undermining social stability. They create forward-thinking policies that reward providers' trustworthy innovation while preventing exploitative practices before harms require intervention.

When these investments get honored, trust accumulates and creates positive feedback loops. People engage more willingly when providers demonstrate genuine value, guided by trust-centered policies, enabling better performance that justifies continued confidence. Providers build sustainable businesses that can fund the continued development of trust-centered, policy-governed technologies for people. Policies achieve their goals of protecting people and enabling beneficial provider innovation by creating frameworks that anticipate challenges rather than merely reacting to problems.

When investments are violated, trust erodes, creating negative feedback loops. People withdraw their engagement and adoption, limiting the system's effectiveness. Providers face reputation damage and regulatory backlash. Policies fail to protect the interests they're designed to serve, creating pressure for restrictive rather than enabling frameworks.

In practice, this dynamic plays out through specific architectural and business model choices.

Venho.AI demonstrates an aligned Trust Exchange. Users invest in the belief that local processing design will genuinely protect their privacy while providing AI assistance. Antti Saarnio and his team of providers invested in local computing and subscription models that eliminate tracking incentives, honoring the trust that people place in privacy-preserving design. European privacy policies reward these business model choices, creating legal advantages for genuine data control. Each party's trust investment is validated: people experience actual control, providers build sustainable businesses on aligned incentives, and policies achieve privacy goals.

Trust accumulates as users discover their confidence in the Venho system was warranted. Greater user engagement provides better feedback for Saarnio's team to improve the system, but without compromising the privacy commitments that earned that engagement. Sustainable subscription revenue supports continued development of trust-centered features. Policy frameworks that reward this approach encourage more providers to adopt similar designs, creating competitive advantages for trust-centered architectures across the industry.

Adobe Firefly reveals how misaligned exchanges violate trust investments. Enterprise customers invested their confidence in Adobe's marketing of "commercially safe" AI, believing their trust in ethical training would be supported by legal defensibility. Adobe invested in marketing campaigns emphasizing superiority over competitors, even as it secretly used the same training practices it had criticized. Regulatory policies that could have created transparency requirements failed to protect people's trust investments from deceptive marketing.

Adobe's undisclosed practices violated the fundamental terms of the exchange. Customers discovered their trust had been exploited for competitive positioning, leaving them with legal liability and reputational risk—precisely what they'd trusted Adobe to prevent. Adobe's competitive claims fell apart when the underlying practices were precisely what they'd criticized. Policy frameworks proved inadequate to protect the trust investments that enable responsible AI adoption, enabling exploitation to persist until a journalist questioned it.

The Trust Exchange framework shows why Trust-Centered Design isn't just ethics—it's economics. When people, providers, and policies align to honor trust investments, the resulting systems can deploy more sophisticated capabilities because users willingly engage when the value is demonstrated, rather than reluctantly accept.

When the Trust Exchange is misaligned around extraction, even technically superior systems face adoption barriers, as potential users abandon the product to protect their trust from exploitation.

This dynamic explains why trust must be designed into AI systems rather than added later. The Trust Exchange happens at the level of business models, user data access, and governance mechanisms. Organizations can't build lasting AI relationships on exploitation-focused foundations, regardless of how compelling their marketing claims become.

Evaluating Trust-Centered AI

Trust-Centered Design becomes actionable when organizations can apply consistent measurement criteria to their AI implementations. The framework provides practical questions that strategic leaders, product teams, and policymakers can use to distinguish between systems that honor trust investments and those that exploit them for corporate gain.

For Strategic Leaders

Strategic decision makers can evaluate AI investments by examining whether proposed systems will strengthen or undermine the Trust Exchange that enables long-term business success.

To what extent does this AI implementation honor the trust investments our stakeholders have made in our organization? Systems that provide genuine value in exchange for user confidence create sustainable competitive advantages, while those that exploit trust for short-term gains face inevitable backlash and regulatory pressure.

How much transparency does this system provide to enable informed decisions, or does it create opacity that demands reluctant acceptance? Trust-centered AI provides users with sufficient understanding of system behavior to make confident choices about engagement levels, rather than demanding trust without providing means to verify.

How well do our business model incentives align with user interests across different contexts and user types? Subscription models, direct payment, and value-based pricing typically align provider success with user satisfaction. In contrast, advertising-based or data monetization models often create systemic pressure toward manipulation rather than service.

For Product Teams

Product developers can apply Trust-Centered Design evaluation criteria at every stage of design and implementation, recognizing that the Trust Exchange varies across user contexts and comfort levels.

What level of meaningful control do users maintain over how their information gets used, and how does this adapt to different user preferences? Trust-centered AI provides granular control over data use, algorithmic behavior, and feature activation, with a design that allows users to calibrate their involvement rather than requiring all-or-nothing consent.

How effectively can users understand system decision making across different use cases, and what design choices enable this understanding? Transparency emerges from experience design, explanation systems, and business model choices that prioritize user autonomy over algorithmic secrecy.

When the system makes mistakes or behaves unexpectedly, what design safeguards provide users with effective recourse? Trust-centered systems embed correction mechanisms, human oversight options, and override capabilities into their core design rather than treating user agency as a trade-off.

For Policymakers

Regulatory frameworks can assess whether they promote conditions that reward trustworthy innovation or enable exploitative practices to proliferate.

To what extent do our requirements foster genuine transparency, or do they enable opaque compliance methods that obscure system behavior? Effective policies focus on user understanding and control rather than technical documentation that serves legal departments but doesn't help people make informed decisions about their level of engagement.

How well do our frameworks anticipate and prevent widespread harms rather than merely reacting after damage occurs? Trust-centered policy design considers the cumulative effects of new technologies across diverse populations and establishes design requirements that enable beneficial innovation while preventing exploitation.

Do our enforcement mechanisms safeguard people's trust investments across contexts, or do they primarily serve providers' interests? Policy frameworks should measure success by user agency and satisfaction across diverse use cases, rather than by industry growth metrics that may conflict with the public interest.

Measurement Criteria

Organizations can measure progress through concrete metrics that reveal whether their design choices serve or exploit user trust.

Transparency That Enables Control: *How effectively can users understand system decision making and predict behavior to calibrate their trust level for their individual circumstances?* Adequate transparency allows users to understand capabilities and limitations rather than encouraging passive acceptance or complete withdrawal.

Incentive Alignment Through Design: *Do the provider's business model and technical design create structural incentives to serve people's interests, and are these incentives supported by policies that reward trust-building over extraction?* Aligned incentives make the Trust Exchange systemic rather than aspirational.

Recourse and Reversibility: *What degree of granular control do users have over system behavior, and can they seek meaningful remedies or exit relationships without losing data or access?* Trust-Centered Design provides mechanisms for users to adjust, contest, or leave as their preferences evolve.

When organizations make Trust-Centered Design a standard evaluation criterion—applied consistently from initial system design through ongoing feature development—they develop organizational capabilities to distinguish between genuine service and sophisticated exploitation. Development decisions create conditions that either enable or hinder the Trust Exchange, making conscious design the primary lever for building systems that honor users' trust investments.

Trust As Foundational

When that mechanic chose the easy route—grab a new coolant cap and charge me for a "missing" one—I learned something fundamental about trust: it's not built through big gestures but destroyed through small deceptions. When teams choose convenience over transparency, or efficiency over honesty, they risk eroding the trust investment that technology depends on.

AI systems face these same micro-decisions thousands of times per day—whether to explain why an algorithm made a recommendation, to prioritize user control over corporate data collection, or to acknowledge uncertainty rather than project false confidence.

Trust-Centered Design creates similar infrastructure for the Mobile Era of Intent. When AI systems operate effectively within the Trust Exchange between people, providers, and policies, individual benefits compound across communities and societies. When users can confidently engage with AI systems because design choices protect their agency, they contribute vital feedback that improves those systems for everyone.

This scales beyond individual adoption. Organizations that build trust-centered AI create sustainable competitive advantages, attracting users who might otherwise abandon the technology to protect themselves from manipulation. When employees can rely on transparent, controllable AI tools, they integrate these capabilities into complex workflows that serve customers more intentionally.

When policy frameworks reward genuine transparency and user control, markets develop competitive advantages around trust rather than extraction.

Trust-Centered Design serves as the foundation for all other pillars of intent-driven AI. Without transparent, controllable systems, Human Connection Enhancement becomes vulnerable to engagement manipulation disguised as relationship building. Anticipatory AI can only serve user goals when the underlying design aligns business incentives with user interests rather than corporate exploitation.

The intimate device relationships that power Mobile as AI Gateway require users to confidently share usage context and preferences, knowing this information serves their intentions rather than corporate surveillance. Even Environmentally Responsible Innovation depends on verifiable sustainability practices, not on marketing claims users must accept blindly.

Without Trust-Centered Design, each pillar becomes vulnerable to exploitation disguised as enhancement. But when trust becomes infrastructure rather than aspiration, AI systems can work together seamlessly across platforms and contexts, creating an integrated experience that finally makes technology serve human intent.

Chapter 7 -

Pillar 3: Seamless Integration

The Interface Never Intended

In 1996, fresh out of college, I joined a small direct marketing agency whose website was embarrassingly simple: a single page of text on a gray background with no HTML table structure.

I volunteered to redesign it. After teaching myself HTML, forms, and database integration, I built a site that was actually quite good by the standards of the day—visually elegant, with clean navigation, an organized content hierarchy, and easy ways to add new pages and capture contact information.

But even as I celebrated that first success, I wondered if my approach was wrong. Every design decision I made was really about training users to think the way our website required them to think. The left-side navigation, the contact forms, and the menu tree structure—all of it needed people to learn our required way of thinking instead of simply expressing what they wanted.

Point-and-click interfaces work well for browsing, but when users arrive with specific goals, they can create friction rather than removing it. This realization deepened over the course of my career as I designed and developed hundreds more websites. I studied usability extensively, became certified in user experience design, and learned the psychological theories behind interface design.

I could cite cognitive load theory, explain why human short-term memory struggles with more than seven items, and justify every design choice with research.

Yet across all those projects, one pattern remained consistent: When I simplified interfaces, user response was better. When I eliminated unnecessary choices, task completion improved. When I reduced the mental effort required to navigate, conversion rates increased—sometimes dramatically.

The data was clear and backed up by studies, but I kept building complex interfaces because that's how the industry made websites.

I got my first hint of what was possible when I watched how Google dominated every other search engine with a simple search box waiting for a keyword. Yahoo, with its category-heavy homepage, was packed with options. Ask Jeeves had a natural language promise but a cluttered interface. Google became the top search engine not just with better algorithms, but by recognizing something I'd been seeing in every usability study: people didn't want a more sophisticated interface—they wanted less.

Today, as I watch conversational AI adoption surge year after year, I'm seeing the same pattern. When technology can understand intent directly, carefully crafted interfaces can become unnecessary translation layers.

To put it simply: the best interface is no interface at all.

That is what is meant by Seamless Integration in the Mobile Era of Intent. That translation layer finally begins to disappear. The goal isn't just better user experience design. It's the coordination of features so seamless that technology fades, finally serving human intent rather than demanding human adaptation.

Defining Seamless Integration

Seamless Integration provides evaluation criteria to distinguish between AI systems that eliminate digital friction and those that create sophisticated cognitive dependency disguised as convenience. This pillar asks a fundamental question: *How does this technology affect users' ability to express their intent directly and maintain control over complex tasks while reducing system complexity?*

The pillar measures progress along three critical dimensions that teams can apply immediately to their AI implementations:

Context Preservation Across Platforms. Current digital systems require users to reconstruct context whenever they move between applications, devices, or services. This dimension measures whether AI systems maintain working context—goals, preferences, recent activities—as users move between different tools rather than forcing them to continually re-establish what they're trying to accomplish.

Cognitive Coordination Without Cognitive Offloading. AI coordination should complement human decision making rather than replace complex reasoning skills. This dimension assesses whether AI coordination reduces overhead while preserving users' understanding of complex processes, or whether convenience gradually displaces the cognitive abilities that enable meaningful contribution.

Interface Dissolution Over Interface Proliferation. Current technology requires users to translate their goals into application-specific commands and navigation patterns. This dimension measures whether AI systems eliminate translation layers by interpreting complex intentions directly and coordinating across multiple tools invisibly.

Seamless Integration requirements vary across different workflows and user sophistication levels. Enterprise systems that manage complex business processes require different integration approaches than consumer applications that support personal productivity. Creative workflows require different continuity mechanisms than analytical tasks. However, all effective integrations serve user intent by enhancing human capability in whatever context users choose to interact.

Strategic leaders can apply this pillar by examining whether proposed AI investments will simplify or complicate the complexity their teams face in achieving multi-step goals. **Product teams** can evaluate whether their systems allow users to express their intent clearly and achieve outcomes more efficiently, while avoiding cognitive dependence. **Policymakers** can assess whether integration standards encourage context preservation and interoperability that enhance user productivity, rather than promoting platform control.

The pillar addresses the misconception that seamless user experiences require surrendering human understanding and control to automated systems. Conversational AI platforms such as ChatGPT, Claude, and Gemini demonstrate both the promise and the peril of Seamless Integration. They achieve interface dissolution by letting users express complex intentions in natural language, and they preserve context across multi-step conversations—but they also risk cognitive offloading by making users dependent on AI for tasks they could otherwise manage themselves. This makes the evaluation framework essential for distinguishing implementations that enhance human capability from those that erode it.

Seamless Integration provides the measurement criteria, but implementation requires conscious choices about interface design and system architecture. Teams that deliberately prioritize intent expression over interface complexity can create the seamless experiences that serve human goals rather than demanding human adaptation.

Design decisions made at the architectural level during development determine whether AI systems become invisible enablers of human capability or sophisticated barriers disguised as innovation.

Conscious Integration

The Jolla Mind2 device, powered by Venho.AI, demonstrates Seamless Integration through a system architecture that prioritizes user workflow continuity and reduces the cognitive effort of context switching across disparate applications. When founder Antti Saarnio designed Venho as an AI operating system from the ground up around human intent, every integration choice reflected a conscious commitment to serving user productivity rather than extracting user data.

Venho's approach to context preservation contrasts sharply with typical AI systems. Rather than forcing users to reconstruct their working context each time they switch between tasks or applications, Venho maintains a vectorized memory system—mathematical relationship mapping—that understands the relationships between projects, conversations, and goals. When a user moves from researching a client proposal to drafting the presentation to scheduling follow-up meetings, the system preserves the contextual thread that connects these activities.

Saarnio's approach to preserving cognitive control is evident in Venho's management of complex workflow coordination. Rather than automatically executing multi-step processes, the system prepares resources and options while preserving user authority over decisions. Users maintain an understanding of how their work is accomplished—they can see the relationships between tasks, understand why specific preparations are warranted, and modify the AI's coordination when circumstances change. This ensures that Seamless Integration enhances human capability rather than creating dependency on automated processes that users cannot inspect or override.

The technical implementation reflects these conscious design priorities. Venho processes information locally on the Mind2 device rather than relying on the cloud for basic functions. This design choice enables genuine orchestration without the AI assuming control—users gain access to AI capabilities without surrendering their data or becoming dependent on external servers for core functionality. The system can coordinate with external services when users choose to do so, but the fundamental intelligence remains under user control.

Saarnio's decision to build Venho around open standards rather than proprietary systems demonstrates conscious integration at the infrastructure level. The system doesn't require users to abandon their existing tools and workflows; instead, it learns to interpret and coordinate with whatever applications and services users already prefer. That coordination represents interface dissolution in practice: Venho reduces the mental overhead of managing multiple systems by understanding user intent and automatically handling coordination.

Venho's machine learning approach reflects these priorities. Rather than training on user behavior to optimize engagement, the system learns preferences and patterns to reduce friction and improve productivity. The AI becomes more helpful without becoming more intrusive.

This conscious integration approach creates what Saarnio describes as "human-centric AI"—technology that adapts to human workflow patterns rather than forcing humans to adapt to technological constraints. When users can express complex intentions through natural interaction and have those intentions translated into coordinated action across multiple systems, technology fades into the background.

The architectural choices underlying Venho reveal how conscious integration serves the trinity of forces that define the Mobile Era of Intent. **Agency** remains intact because users control their data, their processing, and their integration choices. **Trust** develops through transparent local processing that doesn't require the surrender of personal information to external systems. **Intent** serves as a lever that reduces friction between systems, allowing users to accomplish goals without managing digital silos or reconstructing context across platforms.

Most current AI integration follows the opposite pattern, using sophistication to create dependency rather than empowerment. The Jolla Mind2 running the Venho platform demonstrates that conscious design decisions can deliver seamlessly integrated user experiences while preserving user agency—proving you don't have to choose between powerful AI and user control. This either-or thinking stems from business models that profit from cognitive dependency rather than from user empowerment.

Seamless Integration in Practice

The same AI capabilities that can eliminate digital friction and create continuity across complex workflows can also fragment user experiences and introduce unnecessary complexity. Current implementations reveal a spectrum of conscious integration choices, from systems that genuinely reduce cognitive overhead to those that force users to revert to manual processes when AI fails to understand context or intent.

MercuryOS and Desktop Neo illustrate how the dissolution of traditional interfaces can enhance cognitive capability by consciously integrating designs that prioritize human intent over conventional operating-system frameworks. Designer Jason Yuan's MercuryOS replaces standard app containers with flow-based modules that adjust to user intentions.

Similarly, Lennart Ziburski's Desktop Neo redesigns the user experience by eliminating overlapping windows in favor of ephemeral panels and flexible content organization. Both systems minimize the effort needed to manage interfaces without creating cognitive dependency, allowing users to understand their workflow patterns while enjoying seamless interactions. These conceptual designs maintain human decision-making authority by adapting technology to suit human behavior, rather than forcing users to navigate outdated interface hierarchies that fragment attention and require constant mental effort.

Hebbia's AI research platform streamlines complex document-analysis workflows that typically require users to switch between various reading applications, note-taking tools, and synthesis platforms. The system recognizes the relationships between research materials, conversations, and analytical objectives, helping to maintain the user's context as they navigate different types of information. Rather than forcing users to restart their thought process with each new document, Hebbia facilitates a natural progression through research tasks. This allows users to express their intentions conversationally without needing to learn complex database query languages.

Phenom's omnichannel conversational AI maintains context across various platforms, including web applications, WhatsApp, and internal HR portals. This feature eliminates the need for repetitive data entry that often disrupts human resources workflows. Whether employees inquire about benefits via mobile devices or continue the conversation on desktops, the system maintains a complete record of the interaction history and context. This continuity reduces cognitive load for both employees and HR staff, enabling natural conversation flow that facilitates complex administrative tasks without requiring users to navigate multiple applications or repeat information across sessions.

These implementations demonstrate how Seamless Integration design choices can eliminate digital friction while preserving human understanding. Context preservation, cognitive authority, and interface dissolution aren't technical luxuries—they're strategic choices that distinguish genuine productivity enhancement from sophisticated dependency creation.

But the same coordination capabilities that could eliminate translation barriers between human intent and system response can just as easily create algorithmic control disguised as convenience when different technical choices are made.

DPD's AI chatbot showcased how inadequate context handling can turn customer service into a frustrating experience, ultimately damaging brand relationships. Following a faulty update in 2024, the system began to lose track of conversation context, resulting in inappropriate responses, including swearing at customers and labeling DPD as "the worst delivery company in the world." The chatbot's failure to maintain coherent conversation threads forced customers into repetitive loops, requiring them to re-explain their shipping issues multiple times within a single chat. DPD decided to shut down the system on the same day that these failures went viral, reverting completely to human customer service agents. This incident highlighted how AI integration can sometimes increase operational friction instead of reducing it.

Quick-service restaurant AI ordering systems were expected to provide seamless drive-thru experiences, but they ended up creating more friction than traditional human-staffed ordering. In July 2024, McDonald's discontinued its IBM-powered voice AI experiment at more than 100 locations due to widespread order errors, excessive order volume, and numerous customer complaints. Similarly, Taco Bell rolled back its AI ordering initiative at more than 500 locations after encountering issues with contextual understanding. The voice AI struggled with accents and background noise and was susceptible to customer pranks, such as processing ridiculous requests for "18,000 cups of water," which human staff would normally reject. Instead of simplifying the ordering process, these systems required continuous monitoring, corrective protocols, and fallback procedures, leading to more complexity and making simple food ordering more challenging than the human interactions they were intended to replace.

New York City's "Small Business Advisor" chatbot has created more problems than it solved, offering dangerously incorrect legal advice. For example, it suggested that it was legal to fire pregnant employees or serve food contaminated by rats. Users reported in forums that it was faster to call the city than to navigate the chatbot's error-prone responses, which often required constant verification and correction. The system's inability to grasp the context and implications of business regulation advice forced users to double-check every recommendation, resulting in greater cognitive overhead rather than less. As a result, the city had to add disclaimers and retrain the system under human supervision, effectively reinstating the manual oversight that the AI was intended to eliminate. This failure demonstrated that understanding context requires domain knowledge, not just language processing.

The Progress and a Path Forward

These examples demonstrate significant progress and the complexities involved in achieving Seamless Integration. The promise cases illustrate how companies make design choices that maintain user context across platforms, uphold cognitive authority over complex processes, and enable direct expression of intent, rather than relying on an overwhelming array of interfaces.

Even the best-designed solutions highlight the ongoing challenges of Seamless Integration. The catastrophic failures in the peril examples illustrate the importance of having measurement frameworks in place. When teams implement AI without adequate coordination safeguards, or prioritize efficiency over cognitive authority, the outcomes can disrupt user workflows and create more mental strain than the experiences were intended to improve.

The path toward genuine Seamless Integration emerges from conscious design choices that prioritize context preservation over efficiency metrics, cognitive authority over automation, and interface dissolution over feature proliferation. Understanding what this looks like in practice requires examining the fundamental shift in human-computer interaction that enables Seamless Integration.

Evaluating Seamlessly-Integrated AI

Seamless Integration becomes actionable when organizations can apply consistent measurement criteria to their AI implementations. The framework provides practical questions that strategic leaders, product teams, and policymakers can use to distinguish between systems that genuinely reduce digital friction and those that create sophisticated new forms of complexity disguised as convenience.

For Strategic Leaders

Strategic decision makers can evaluate AI investments by examining whether proposed systems will preserve context across workflows, maintain decision-making capability, and reduce coordination complexity without creating overly complex dependencies.

How does this AI system handle context preservation when our teams move between different projects, departments, or external collaborations? Effective Seamless Integration maintains organizational knowledge and workflow continuity across disparate platforms. Systems that lose context when users switch between different work streams or require constant re-explanation of organizational priorities fail to deliver genuine integration benefits.

How can this AI system maintain users' understanding of complex processes while reducing coordination complexity, ensuring that our teams retain cognitive authority rather than gradually surrendering decision making and problem solving to automation? Organizations should measure whether Seamless Integration preserves human agency while enhancing productivity, preventing the gradual erosion of institutional knowledge disguised as operational efficiency.

What evidence is there that this AI implementation enables the intuitive expression of complex intentions, rather than requiring our teams to learn new technical interfaces or system-specific commands? Effective interface dissolution leads to measurable productivity gains by reducing context switching and training overhead. In contrast, unsuccessful implementations often necessitate additional support resources and provoke user resistance, undermining adoption goals.

For Product Teams

Product developers can apply Seamless Integration criteria during design and testing phases to ensure AI features eliminate rather than create digital friction.

Does our conversational AI maintain context across extended user sessions and complex workflows without requiring users to restart or re-explain their goals? Practical context preservation requires a technical design that accounts for the relationships among user intentions, project histories, and workflow patterns. Systems that disrupt the conversation or force users to manage context manually create additional cognitive demands rather than alleviating them.

How does our AI system help users develop better workflow judgment and situational awareness over time, rather than handling complexity without user understanding? Technical implementations should ensure that users understand how work is done while benefiting from coordination assistance and avoiding practical dependency, which can be misleadingly presented as helpful automation that obscures decision-making processes.

Can users understand and control how our AI system makes decisions that affect their workflows, and do they maintain meaningful ability to modify system behavior when it doesn't align with their intentions? Seamless Integration preserves user control by demonstrating how coordination works, enabling users to communicate goals naturally, and maintaining control over results.

For Policymakers

Policy frameworks can encourage Seamless Integration that serves public interest while preventing exploitation of AI capabilities for market concentration or user manipulation.

Do our regulatory approaches encourage AI implementations that preserve workflow continuity and context preservation across different platforms and services without forcing users into proprietary ecosystems? Policy frameworks should measure success by user agency and operational continuity rather than industry growth metrics that may conflict with genuine productivity enhancement.

How can regulatory frameworks ensure that AI integration enhances rather than replaces human cognitive capacity and decision-making skills, particularly in highly regulated domains? Policy approaches should distinguish between AI that helps people think better and systems that gradually replace human learning and expertise with automation.

How effectively do our standards enable intuitive intent expression that serves public interest rather than optimizing for engagement metrics that may conflict with user productivity? Policies should distinguish between AI systems that enhance human capability and those that exploit human behavior for commercial advantage, with clear standards for measuring genuine public benefit.

Measurement Criteria

Organizations can measure progress through concrete metrics that reveal whether design choices serve genuine Seamless Integration or create sophisticated dependency.

Context Preservation Across Platforms: *How effectively do systems maintain user working context—goals, preferences, recent activities, and relevant information—as users move between different tools and environments without requiring manual reconstruction?* Effective integration maintains the continuity of user intent across complex operational boundaries, providing a seamless experience instead of requiring users to repeatedly reaffirm their intent.

Cognitive Coordination Without Cognitive Offloading: *Do users maintain comprehension of complex processes while benefiting from AI coordination, retaining cognitive authority and decision-making capacity without becoming solely reliant on AI systems?* Genuine Seamless Integration enhances human decision-making capacity while reducing coordination overhead, ensuring that users understand how work is accomplished rather than simply delegating cognitive tasks for the sake of automated convenience.

Interface Dissolution Over Interface Proliferation: *Are users able to accomplish complex goals through intuitive expression of intent, reducing the mental burden of navigating digital tools, or do they encounter new layers of interaction complexity disguised as convenience?* Actual interface dissolution allows direct communication of multi-step intentions without requiring users to manage AI behavior, optimize prompts, or issue system-specific commands.

When organizations make Seamless Integration a standard evaluation criterion—applied consistently from initial AI system design through ongoing feature development—they prevent the gradual erosion of human cognitive skills disguised as innovation.

Technology Served Seamlessly

Three decades after creating my first website, which required users to align with our way of thinking, I now understand what I once sensed but couldn't articulate. The user interface serves as a translation layer between human intent and machine capability. The more sophisticated the interface becomes, the more detached human intent is from achieving the actual goal.

As conversational AI adoption continues to grow, technology stands to eliminate the translation work. Users can express complex multi-step intentions directly, and systems can coordinate responses across multiple capabilities without requiring users to learn platform-specific commands or navigate hierarchical menus.

The Seamless Integration pillar measures whether AI systems truly dissolve interface complexity or merely create more sophisticated forms of it. The distinction matters because the same technological capabilities that could eliminate digital friction can also perfect cognitive offloading—making dependency feel like convenience while gradually eroding critical problem-solving skills and human decision-making capacity.

Most current integration solutions replace human decision making with algorithmic control, creating overly complex systems that gradually erode the added value of human comprehension while appearing to provide convenience. True Seamless Integration requires conscious design choices that preserve human understanding of complex processes while handling coordination—enabling human capability rather than replacing human cognition with automation.

When teams consistently apply the Seamless Integration evaluation criteria, they eliminate translation layers between human intent and system capability, enabling the direct expression of complex goals without interface mediation. Organizations that master this dissolution of complex interfaces create conditions in which technology fades into the background—not because it's invisible, but because it removes the friction between intention and execution that traditional user experiences impose.

When AI systems truly understand workflow context and coordinate across platforms without requiring interface management, they begin to recognize patterns in how human work actually gets accomplished. This creates conditions in which technology can start to anticipate needs rather than wait for explicit instructions, while also eliminating friction between intent and execution rather than substituting AI efficiency for human judgment.

Chapter 8 -

Pillar 4: Anticipatory AI

The Information I Knew Existed

It was a regular Tuesday when I reached for my phone and felt nothing. That sudden absence created an immediate wave of anxiety. My iPhone isn't just a device—it's my external brain, my connection to work, family, everything.

I did what many people now do reflexively—raised my wrist and spoke into my Apple Watch. "Find my iPhone." The familiar ping echoed from my bedroom. Problem solved.

Except it wasn't.

I didn't need to call anyone or check messages. What I needed was specific information buried somewhere in my phone—words I'd written to someone, though I couldn't remember who or when. Just the content. I knew it existed somewhere.

Standing there, holding more computing power than was carried on the Apollo 11 spacecraft, I found myself manually hunting through dozens of apps, email threads, text conversations, and document folders. Searching for information that I was certain existed, but couldn't locate.

The thing was that Siri had flawlessly executed a reactive command from me to find my lost device, but even if I'd explained exactly what I needed—"Find the words I wrote about marketing technology"—it had no capacity to understand or address my actual intent.

This was late 2022. ChatGPT had just entered public consciousness, and everyone was talking about what "artificial intelligence" could do. Yet here I was, experiencing the fundamental gap between technology that responds to explicit commands and systems that could understand context and anticipate human needs based on intent.

The friction felt unnecessary because I knew what was missing. The information was scattered across my apps and accounts, but Siri couldn't connect my intent to my data. I wondered: surely somebody was working on this problem. I expected announcements any day about assistants that could understand "find my marketing technology notes" and actually deliver results.

But those announcements never came. Instead, a powerful new AI capability had arrived primarily without a clear problem statement, leaving companies, users, and even developers scrambling to figure out what it was actually for. The conversation seemed focused on "what can we do with this technology?" rather than "what human problems should this solve?" The speed of deployment—the rush to define the technology's purpose—outpaced any deliberate debate to ensure it served genuine human needs.

Standing in my doorway, phone in hand, manually scrolling through app after app with no way for my device to understand what I actually needed, I realized this was the moment our relationship with technology could fundamentally change. The experience highlighted a crucial disconnect: I had clear intent, my phone had all the information, but there was no bridge between human need and system capability.

That's the promise of Anticipatory AI in the Mobile Era of Intent—systems that understand context well enough to predict what information you need, when you need it, and deliver it proactively before you have to manually hunt through disconnected apps for information you know exists.

Defining Anticipatory AI

Anticipatory AI provides evaluation criteria to distinguish between AI systems that predict user needs to accomplish goals and those that predict behavior to manipulate outcomes. This pillar asks a fundamental question: *How does this technology anticipate and respond to user goals, and how do those interactions affect human decision-making capacity and autonomy?*

The pillar measures progress along three critical dimensions that teams can apply immediately to their AI implementations:

Contextual Over Behavioral. Many current AI systems track behavioral patterns to manipulate user engagement rather than understanding what users actually want to accomplish. This dimension assesses whether AI learns to recognize when users need specific support to achieve their goals, or whether it learns to use psychological vulnerabilities for behavioral control.

Proactive Not Presumptuous. Effective anticipatory systems surface resources and options while preserving human authority over decisions. This dimension evaluates whether AI actively gathers information and suggests actions without making assumptions about preferences or automating choices that should remain human.

Transparency Over Opacity. Anticipatory capabilities should operate with reasoning that users can interpret, evaluate, and correct when necessary. This dimension assesses whether users understand why systems anticipate particular needs and maintain meaningful control over predictions, or become dependent on algorithmic assumptions they cannot guide.

The evaluation framework recognizes that anticipatory requirements vary across different domains and user sophistication levels. Healthcare AI that anticipates patient needs requires different transparency and oversight mechanisms than productivity AI managing professional workflows. Creative work demands different anticipatory support than analytical task completion. However, all effective Anticipatory AI systems must enhance human capability while preserving meaningful control over significant decisions.

Strategic leaders can evaluate whether proposed AI investments will enhance their teams' decision-making capacity or create automated dependency disguised as efficiency gains. **Product teams** can evaluate whether their anticipatory features enhance user capability by understanding contextual goals and preserving human authority over decision making, or whether they create sophisticated dependency through behavior tracking and overindulgent automation. **Policymakers** can assess whether

Anticipatory AI systems provide sufficient transparency for regulatory oversight and user comprehension, or whether opaque predictive solutions create accountability gaps that enable systemic exploitation without recourse.

Anticipatory AI provides measurement criteria for distinguishing between systems that understand user intent well enough to prepare without presuming. This evaluation framework becomes essential as AI capabilities advance toward systems that predict needs with increasing accuracy. The same technologies that could eliminate routine mental effort and enable humans to focus on complex strategic thinking could also perfect behavioral manipulation, making conscious anticipation increasingly difficult to distinguish from algorithmic control.

Conscious Anticipation

Venho.AI's approach to Anticipatory AI reveals the difference between systems that predict user engagement and those that understand user context to provide genuine assistance. When Antti Saarnio rebuilt Venho's architecture in 2025, spending seven months working 100-hour weeks to completely re-architect the system, he made a fundamental choice that defines conscious anticipation: the system learns workflow patterns to reduce friction, not to optimize for platform engagement at the expense of user autonomy.

This design decision emerged from Saarnio's realization that "Spectra Models are not really going to create a memory for people." These open-source LLMs, designed to address memory and bandwidth limitations, make it harder to identify what users need rather than solving the challenge of contextual assistance.

Venho addresses this through what Saarnio calls "context engineering" rather than "prompt engineering." Venho splits each assistant into specialized modes—email, calendar, task, travel, and health—that users can define and customize. This narrowed contextual approach allows the system to focus on relevant information patterns rather than building broad behavioral profiles. What makes this truly anticipatory is how Venho takes what cognitive scientist Göte Nyman calls a "personal perspective to data"—organizing information not according to database logic, but according to how individual users think about their work, relationships, and priorities.

When working on a presentation, Venho follows a structured workflow: understanding user intent, checking for relevant information and existing templates, then presenting options while preserving human authority over decisions. The AI coordinates across thirty integrated tools, but does so by preparing resources rather than automating decisions. Crucially, what Saarnio calls "deterministic software" ensures that the AI actually follows user intentions rather than optimizing for its own operational efficiency.

This represents genuine anticipatory assistance rather than sophisticated automation disguised as helpfulness. While some other AI systems track user behavior to predict actions that serve platform objectives, Venho's local processing design makes this data-harvesting approach technically impossible. There's no central server collecting behavioral data for automated manipulation. Instead, the system develops what Nyman describes as "intelligently organized memory data" that lives on your device and serves your productivity goals. As Saarnio explains, "Human in control at every significant phase of AI actions"—the system suggests and prepares, but users retain autonomy over final decisions.

This creates what Saarnio calls a "human-centric AI," as opposed to the big-tech approach of "resource-centric AI," in which people are essentially resources in service of the system. The efficiency of Venho's 3B (3-billion) parameter model enables local processing that becomes more cost-effective than cloud usage fees and more private than remote data processing. In contrast, the subscription model aligns business goals with user satisfaction rather than engagement metrics that may conflict with user well-being.

The architectural decisions underlying Venho demonstrate how Anticipatory AI can serve the trinity of forces that define the Mobile Era of Intent. **Agency** remains intact because users control their data processing, workflow customization, and integration choices. **Trust** develops through transparent local processing that doesn't require surrendering personal information to external cloud systems. **Intent** serves as the organizing principle that enables AI to predict and prepare without autonomous decision making, creating technology that understands what users want to accomplish rather than manipulating what they do.

This represents the fundamental choice between conscious anticipation, which amplifies human capability, and overindulgent automation that replaces human judgment.

Anticipatory AI in Practice

Anticipatory AI becomes clearer when examining how different AI implementations either prepare intelligently to serve human goals or predict human behavior to manipulate it toward algorithmic control. Recent deployments reveal distinct contrasts between systems designed to enhance human decision-making capacity and those optimized to replace human judgment with automated efficiency.

When organizations face decisions about predictive capabilities, contextual awareness, and user agency, their development decisions either strengthen or undermine the conscious anticipation that genuine productivity partnerships require.

TREWS is a sepsis detection system that employs Anticipatory AI to enhance clinical judgment in critical situations rather than replace it. This system has been implemented across five hospitals in the Washington D.C. metropolitan area, including Johns Hopkins, and monitors nearly 600,000 patients. It identifies patterns indicating sepsis risk and issues alerts to clinicians for evaluation, supporting rather than substituting for medical expertise. When providers respond to alerts within three hours, patients experience a 18.7% reduction in hospital mortality rates and shorter lengths of stay. Notably, clinicians assessed 89% of the alerts generated but confirmed only 38% as actionable. This discrepancy highlights the high stakes associated with early sepsis detection and the challenges in identifying this life-threatening condition in time to improve patient outcomes.

GitHub Copilot's code completion feature exemplifies Anticipatory AI, prioritizing programmer control over automation efficiency. This developer tool analyzes the code's context to predict what programmers might need next. However, it requires the user's explicit acceptance—either by pressing the Tab key to view suggestions or by choosing alternatives. This seemingly simple design choice reflects a fundamental philosophy in user experience design: offer options without making decisions on the user's behalf. Research involving 95 professional developers found that those using Copilot completed JavaScript tasks 55.8% faster while maintaining full control over their code decisions. Developers reported higher satisfaction because they retained authority over their work.

AI-supported mammography screening has shown that anticipatory assistance can enhance patient outcomes while maintaining radiologists' authority. In trials involving over 80,000 women, AI technology triaged exams and flagged suspicious cases, allowing radiologists to retain control over diagnostic decisions. The rate of cancer detection remained comparable to that of screenings conducted solely by humans, and radiologists' workload decreased by 44.3% without an increase in false positives. When this approach was implemented nationwide in Germany, it led to a 17.6% increase in detection rate by managing obvious and typical cases, prompting radiologists to review only genuinely suspicious findings.

These implementations show how Anticipatory AI design can improve human decision making by actively providing intelligent options. Contextual learning, proactive assistance, and transparent reasoning are not merely features to be added later—they are intentional choices that differentiate true empowerment from manipulation of user behavior.

However, the same anticipatory capabilities that can predict user needs to accomplish goals can just as easily make presumptuous decisions that override human choice when different business models and design priorities prevail.

Microsoft Viva Insights is an automated performance management tool that presents Anticipatory AI as productivity assistance, but obscures its potential for advanced surveillance of workers. This system analyzes patterns in calendars, email response times, and collaboration metrics to produce "productivity scores" and automated suggestions. However, these suggestions often prioritize efficiency metrics over the actual quality of work or the well-being of individuals. A study of knowledge workers in higher education found that many felt monitored and pressured to change their work habits to improve their system ratings, rather than to focus on meaningful productivity. The AI replaces human judgment regarding work-life balance with corporate oversight objectives, creating a form of anticipatory surveillance that ultimately benefits management more than it benefits workers' well-being.

Meta's suicide prevention AI highlights the risks associated with anticipatory systems designed to detect mental health crises. While these systems have been in place since 2017, using machine learning to identify suicide risk signals in posts and livestreams, the intervention operates without proper regulation. By 2018, they reported that they had alerted human reviewers and emergency responders to more than 3,500 cases each month. However, the technology lacks necessary medical and mental health regulatory oversight, with no publicly available data on its effectiveness, no independent validation, and no clear protocols that adhere to clinical ethics standards. Without these elements, the system risks providing unregulated interventions rather than genuine mental health support.

Uber's surge-pricing algorithms demonstrate a form of anticipatory abuse through unclear, automated processes. These systems predict demand patterns that do not reflect user intentions, instead creating complex pricing dynamics that primarily maximize corporate revenue at the expense of passengers. The algorithms produce unpredictable fare multipliers for riders while simultaneously affecting driver earnings through obscure adjustments. In the UK, drivers have reported fare reductions of up to 25% due to arbitrary manipulations that passengers cannot control. Rather than genuinely addressing transportation needs, these anticipatory capabilities exploit information imbalances to increase corporate profits. Transparent pricing algorithms could better predict demand while ensuring fair and predictable rates for both passengers and drivers.

The Distinguishing Design

These examples show that technological capabilities enabling intelligent preparation can also manipulate user behavior and outcomes, disguised as helpful assistance. The critical difference lies in whether anticipatory systems are designed to empower human decision making or to predict behavior to normalize corporate control.

Anticipatory AI that serves human intent provides transparency into its predictions, preserves human agency over actions, and measures success by user satisfaction with outcomes rather than by measuring feature engagement. Systems that exploit anticipatory capabilities operate through opacity, measure success through behavioral tactics, and gradually replace human judgment with computer logic.

When teams consistently apply Anticipatory AI evaluation criteria—contextual learning over behavioral tracking, proactive preparation over presumptuous automation, transparent reasoning over opaque algorithms—they create competitive advantages by improving human capabilities instead of fostering complex dependencies.

Evaluating Anticipatory AI

Anticipatory AI becomes actionable when organizations consistently apply measurement criteria to distinguish between systems that predict user needs to accomplish goals and those that predict user behavior to manipulate outcomes. The framework provides practical questions that strategic leaders, product teams, and policymakers can use to evaluate whether AI implementations serve conscious anticipation.

For Strategic Leaders

Strategic decision makers can evaluate AI investments by examining whether proposed systems will enhance human decision-making capacity through intelligent preparation rather than by manipulating behavior.

How can this AI system learn our teams' work patterns and life rhythms to become more helpful while protecting against psychological manipulation? Systems should recognize individual context and goals to enable conscious anticipation without exploiting personal behavior for automated surveillance.

How can this technology surface resources and coordinate preparation without making decisions on our behalf while ensuring human decision making remains meaningful and accessible? Implementations should enhance human decision making while ensuring cognitive authority and avoiding reliance on automation.

How can this system help our teams develop better judgment and situational awareness while maintaining appropriate human oversight of AI coordination? Organizations should maintain institutional knowledge and enhance cognitive capacity while utilizing the benefits of Anticipatory AI assistance.

For Product Teams

Product developers can apply Anticipatory AI evaluation criteria during the design and testing phases to ensure that features provide conscious assistance rather than extraneous and manipulative automation.

How effectively does our Anticipatory AI learn user context and goals to provide increasingly helpful support, rather than tracking behavior patterns that could enable manipulation? Product teams should assess whether their systems understand user goals and become increasingly supportive of those objectives over time, instead of exploiting psychological vulnerabilities or response patterns that prioritize system goals.

How can our anticipatory features actively prepare options and coordinate resources while preserving user authority over decisions and maintaining meaningful human oversight? Intelligent anticipatory assistance should present relevant information and recommend suitable actions while ensuring users retain control over their choices.

How can our system help users develop better workflow judgment and situational awareness while providing sophisticated coordination support and preventing practical dependency? Anticipatory AI systems should enhance user understanding of how tasks are accomplished while providing efficiency, productivity, and convenience.

For Policymakers

Policy frameworks can encourage conscious anticipation by creating incentives for contextual understanding, intelligent assistance, and user empowerment, rather than anticipatory exploitation.

How can regulations distinguish between AI systems that understand user context and those that exploit behavior patterns while encouraging both innovation and user protection? Policy frameworks should promote intelligent assistance that aligns with users' goals while limiting manipulative practices that benefit corporations at users' expense.

How can requirements ensure AI systems preserve meaningful human control over consequential decisions while enabling helpful automation? Effective oversight should enhance human decision making rather than replace human judgment with computer logic—especially in employment, housing, lending, and health outcomes.

How can data protection laws ensure that anticipatory capabilities strengthen user decision-making capacity while enabling contextual learning? Regulatory approaches should ensure transparency regarding how AI systems learn user patterns and maintain user control over the application of those insights.

Measurement Criteria

Organizations can measure progress through concrete metrics that reveal whether design choices serve conscious anticipation or create sophisticated manipulation.

Contextual Over Behavioral Learning: *Do users report that the system becomes more helpful over time by understanding their work patterns and goals, or do they feel manipulated by automated responses to their usage patterns?* Anticipating user needs leads to greater satisfaction with outcomes, while manipulating behavior fosters engagement without empowering users.

Proactive Over Presumptive Assistance: *Do users maintain authority over essential decisions while the AI prepares helpful options, or does the system gradually take over decision making as it handles more complex tasks?* Genuine assistance enhances human capabilities and ensures user understanding, while assumptive systems create dependency by disguising themselves as helpful automation.

Transparent Accountability Over Opaque Systems: *Can users understand why the system anticipates particular needs and maintain meaningful control over how those insights get applied, or do they become dependent on system processes they cannot evaluate or correct?* Transparent anticipatory AI fosters trust through clear reasoning, while opaque systems enable manipulation through prediction methods that users cannot evaluate or contest.

When organizations standardize Anticipatory AI as an evaluation criterion—applied consistently across three dimensions—they develop the capability to differentiate between genuine assistance and algorithmic control that masquerades as helpful technology.

Intelligence That Finally Anticipates

Standing in my hallway three years ago, manually hunting through dozens of apps for content I knew existed somewhere in my digital ecosystem, I experienced the gap between reactive commands and anticipatory understanding. Siri could find my lost device when explicitly asked. Still, no system could bridge the space between "I need something I wrote about marketing technology" and the actual location of that information across my fragmented digital domain.

That friction felt unnecessary because I could envision what should be possible. The information existed across my digital ecosystem—every word I'd typed, every document I'd created, every conversation I'd had was stored somewhere in the cloud, across various platforms and services. But even with unified access, understanding intent represents a fundamentally different technological challenge from executing commands. This is precisely the gap that Anticipatory AI is designed to close. True Anticipatory AI must comprehend human goals directly rather than requiring users to translate their needs into system-compatible queries.

The Anticipatory AI pillar provides evaluation criteria for navigating these complex trade-offs. It helps teams assess whether AI systems primarily predict user needs to enhance human decision making or instead predict user behavior in ways that may compromise user agency. The difference matters because most systems fall somewhere along this spectrum. The same technological capabilities that could eliminate routine cognitive overhead can gradually drift toward behavioral manipulation, making dependency feel like intelligent assistance. Teams need frameworks for recognizing when anticipatory features serve genuine user needs rather than optimizing for metrics that may conflict with user well-being.

But Anticipatory AI cannot navigate without coordination with the other pillars. Human Connection Enhancement ensures that predictive capabilities reduce the digital overhead that interferes with relationships and meaningful work. Trust-Centered Design provides the architectural transparency and user control that enable confident engagement with prediction-based systems rather than requiring passive acceptance of algorithmic decision making. Seamless Integration coordinates anticipatory capabilities across platforms and contexts without creating new dependencies or forcing users to manage complex AI behavior across multiple systems.

Without conscious anticipation grounded in contextual understanding, proactive assistance, and the preservation of agency, Anticipatory AI transforms into sophisticated behavioral manipulation masquerading as beneficial technology. But when proactive systems truly understand human intent, they create the foundation for technology that finally serves human capability rather than replacing human judgment.

This progression—from reactive commands to contextual understanding to anticipatory assistance—represents more than technological advancement. It signals the emergence of human-centered AI: a fundamental shift in which human intent becomes the primary interface for digital capabilities, making technology finally serve the goals we are trying to accomplish rather than the commands we must remember to execute.

Chapter 9 -

Pillar 5: Mobile as AI Gateway

Intimate Gateways

The checkout line at the grocery store moved quickly, but when I got to the front, I realized my wallet wasn't in my pocket. Instead of the familiar leather rectangle, I found lint.

Ten years ago, this would have been mortifying. I remember when this happened to my wife, with kids in tow and a cart full of groceries, having to explain to the clerk that she'd forgotten her wallet and would need to put everything back. The embarrassment, the frustrated drive home, wondering what that person thought, even though they were so kind about it.

This time? I held my phone next to the card reader. Apple Pay completed the transaction before the cashier even looked up—problem solved in seconds, no awkwardness, no disruption. Forgetting my wallet was so inconsequential that it barely registered as an inconvenience.

But forgetting my phone? Regardless of the situation, that would send me racing back home, no matter how far I'd driven.

The difference emphasizes how much my relationship with mobile technology has evolved. My phone has transformed into a gateway, offering features once reserved for separate devices—payment systems, personal identification, navigation services, and unlimited access to information. Beyond mere convenience, mobile devices are becoming the primary gateway to digital capabilities worldwide.

Consider the scope: 57% of the world's population now uses mobile internet on their own devices, with mobile devices accounting for most global web traffic. In many regions, mobile has leapfrogged traditional infrastructure entirely—40% of adults in Sub-Saharan Africa use mobile money accounts, even in areas where bank branches have never existed. In comparison, 15% of U.S. adults rely exclusively on smartphones for internet access.

Yet 43% of humanity remains entirely outside the mobile internet gateway—either without coverage or unable to access available networks.

This isn't just about communication anymore. Mobile devices have become the primary interface between humans and digital capabilities for most of the world's population, and now artificial intelligence is converging on this same platform. GenAI-capable smartphones are projected to reach 400 million shipments in 2025, as processors become increasingly capable of running sophisticated AI models locally without cloud connectivity.

The implications are staggering. The same device that replaced my wallet is poised to unlock AI capabilities that currently require cloud computing, enterprise software, or specialized expertise. This represents unprecedented democratization of sophisticated AI capabilities at a global scale.

But what concerns me most is who will ultimately control these gateways? My Apple Pay transaction was convenient, but it also created new dependencies and potential vulnerabilities. As mobile devices increasingly serve as the primary interface for AI capabilities, every design decision regarding privacy, data security, access controls, and algorithmic transparency becomes a conscious choice about how this technology can democratize opportunity and enhance human capability.

What began as communication devices has become gateways to capabilities that were unimaginable just a decade ago. The trajectory is accelerating, with AI processing moving directly onto phones and decision making shifting from cloud servers to devices in our pockets. Each implementation choice—about privacy, about access, about who controls the underlying models—shapes how these intimate gateways can serve human intent and enhance individual capability through contextual understanding.

Defining the Mobile AI Gateway

Mobile as AI Gateway provides evaluation criteria to assess how effectively AI systems leverage mobile's democratizing potential while preserving user agency and contextual privacy. This pillar asks a fundamental question: *How effectively does this mobile AI implementation harness the democratizing potential of ubiquitous, intimate technology while preserving human agency and serving genuine intent?*

The pillar measures progress along three critical dimensions that teams can apply to their AI implementations:

Personal Intelligence Sophistication. Current mobile AI relies primarily on cloud-based processing and general-purpose models rather than on understanding individual patterns and contexts. This dimension assesses whether AI develops capabilities appropriately sized for mobile platforms and adapts to individual working styles and communication patterns.

Universal Access Architecture. Mobile's global reach creates unprecedented potential for democratizing advanced AI capabilities, but universal access requires architectural choices that prioritize broad accessibility over optimization of premium features. This dimension evaluates whether sophisticated AI assistance becomes available across diverse device capabilities and economic circumstances, or remains concentrated among users who can afford cutting-edge hardware and unlimited data access.

Contextual Intimacy Stewardship. Mobile devices continuously observe personal, behavioral, and contextual information—more than any technology in history. This dimension assesses how responsibly AI systems handle access, whether intimate contextual knowledge enables genuine empowerment through personalized assistance or becomes a means of behavioral manipulation and systematic data extraction.

People's relationships with mobile technology are marked by complex dynamics of dependence and distrust, as well as convenience and concern. Although mobile devices provide unprecedented access to personal information and behavior, users often feel conflicted about their reliance on these technologies, even as they do not fully trust them. Implementations of mobile AI vary in terms of sophistication, accessibility, and responsibility, existing along a spectrum rather than fitting into strict binary categories.

Strategic leaders can apply this pillar by assessing how mobile AI investments enhance individual capability while preserving user agency, and how implementations can democratize access across diverse user contexts and technical abilities. **Product teams** can evaluate how effectively their systems provide genuine personalization that enhances user capability through contextual understanding while preserving privacy and agency. **Policymakers** can consider how frameworks can encourage implementations that democratize AI access while ensuring intimate contextual relationships provide proportional user benefit.

This evaluation criterion becomes essential as AI capabilities scale across devices that billions of people carry everywhere and rely on for crucial functions. The choices made now about mobile AI architecture, accessibility, and stewardship determine whether this convergence empowers human flourishing or creates the most sophisticated manipulation apparatus in history, disguised as personal assistance.

Understanding how these principles translate into practice requires examining systems that prioritize democratization over extraction, beginning with architectures designed around user agency.

Conscious Gateway

Venho.AI demonstrates Mobile as AI Gateway through conscious design choices that prioritize distributed intelligence over centralized control. When Antti Saarnio decided how Venho would scale across devices, he faced fundamental questions about where AI should reside and how users should access it: Should intelligence reside in corporate clouds, with devices as "dumb terminals," or should users control where their AI processing occurs while maintaining seamless mobile access?

The contrast with conventional mobile AI becomes evident in Saarnio's critique of how platforms treat mobile technology as a gateway. When Apple announced its Apple Intelligence capabilities, Saarnio pointed out a more significant issue beyond technical capabilities. While companies claim they are developing personal AI, their infrastructure choices indicate different priorities. "What I know currently, we don't have the data infrastructure in place to handle that on the phone," Saarnio observed. His concern wasn't technical capability—it was architectural philosophy that prioritizes platform control over user autonomy.

Venho's architecture addresses this gap by embracing distributed processing while preserving user control. Saarnio implements what he calls a "node model" architecture: user memory is processed in a single trusted location—either a home server or a cloud location the user explicitly controls—while thin memory layers synchronize with devices. This approach acknowledges that mobile devices could serve as interfaces to intelligence processed elsewhere, while ensuring that users determine where that processing occurs.

Saarnio explains that "distributed memory processing" represents the architectural path forward for mobile AI that serves human intent. His approach recognizes that mobile devices will increasingly serve as interfaces for AI processing while preserving user sovereignty over where that intelligence resides. "People have become so dependent on [their screens]," Saarnio observes, which makes the question of control over AI processing strategic rather than merely technical.

This control consideration fundamentally shapes Venho's Mobile as AI Gateway philosophy. As Saarnio discusses with enterprise clients, organizations increasingly treat AI "like cloud access"—and just as they learned to plan for internet outages, they need strategies for when AI connectivity fails. "We are talking about mission-critical systems, basically, that everything stops in the future if your AI stops when everything is integrated." The node architecture addresses this vulnerability by maintaining essential intelligence locally while coordinating with more powerful external processing that users control.

This approach also entails leveraging existing computational resources rather than requiring new infrastructure. When Saarnio advises industrial clients, he asks: "Why would you acquire huge computing capacity if you have desktops already, which can process a third of the computing for the AI?" The architecture distributes processing intelligently by handling what it can locally, coordinating with the user's trusted server when needed, and accessing cloud resources only when directed.

The technical implementation reflects these conscious design priorities. Venho is a new way to interact with technology. As Saarnio describes it,

"I don't have to open any application ever anymore, basically." The Jolla Mind2 device serves as the server node in this architecture—users download Venho to their desktop or mobile devices. They can then purchase a Mind2 as the trusted processing hub that coordinates their personal AI ecosystem.

The technical decisions underlying Venho's gateway approach reveal how mobile AI can serve the trinity of forces that define the Mobile Era of Intent. **Agency** remains intact because users control where their intelligence lives and how devices access it. **Trust** develops through transparent distributed processing that doesn't require surrendering personal data to platforms optimizing for hidden objectives. **Intent** becomes the organizing principle—the mobile device understands what users want to accomplish and coordinates access to AI capabilities needed to serve those goals, whether processing happens locally, on a trusted server, or through cloud services.

The approach creates "human-centric AI" that adapts to how people actually use mobile devices rather than forcing users to adapt to technological constraints. Mobile becomes the gateway not because it's the most powerful computing platform, but because it's the most personal and accessible interface to AI capabilities. The device people carry everywhere becomes the trusted gateway to intelligence that lives within their control, whether in their pocket, on their desk, or in the infrastructure they choose to trust.

This represents the fundamental choice between mobile as a genuine gateway to user-controlled AI versus mobile as a surveillance portal to platform-controlled intelligence. When the architecture serves users rather than platforms, Mobile as AI Gateway enables access without demanding surrender.

Mobile AI In Practice

The same technological capabilities that could democratize sophisticated AI assistance through mobile devices can just as easily concentrate that intelligence behind corporate gatekeepers who control access, extract behavioral data, and determine whose intent gets served. Current mobile

AI implementations reveal a spectrum of conscious architectural choices, from systems designed around user sovereignty to those that exploit intimate access for corporate surveillance.

Recent deployments reveal sharp contrasts between teams that prioritize distributed intelligence for user benefit, and those that optimize mobile devices for extraction and control. When organizations face fundamental decisions about where AI processing occurs, who controls the models, and how contextual data is handled, their architectural choices either honor or violate the trust investment that Mobile as AI Gateway represents.

MLC Chat demonstrates distributed intelligence through infrastructure decisions that prioritize user sovereignty over platform control. Built on the MLC runtime, this iOS and Android application runs smaller, open-weight models—Llama 2, Mistral, Gemma—entirely on-device, functioning fully offline once downloaded. Users select their appropriate size and weight models while keeping prompts and context entirely local, with no server communication. This approach proves that mobile AI can serve personal intelligence without requiring users to surrender conversational data to platform providers.

Aiko and Whisper Notes preserve intimate context through privacy-by-design architecture. These iOS apps process voice recordings using OpenAI Whisper models, running entirely offline on the user's device, ensuring that sensitive audio—personal notes, interviews, medical dictations—never leaves the device. Developer documentation confirms "all transcription happens locally; no audio leaves your phone." This implementation approach demonstrates how AI can listen to users rather than at them, transforming intimate contextual access into personal empowerment. Rather than corporate intelligence-gathering, this localized approach enables sophisticated language understanding with complete user control.

Google LiteRT and Qualcomm's on-device capabilities show the architectural foundation is evolving toward genuine personal AI. These frameworks enable incremental model training and personalization directly on mobile devices, moving beyond static inference toward AI that learns privately from individual users. LiteRT demonstrates fine-tuning models within Android applications, while Qualcomm's NPU capabilities enable accent adaptation without sharing data externally. Together, these frameworks reveal the technical groundwork for mobile AI that genuinely resides with users—adapting to individual patterns while maintaining complete privacy through local processing.

These implementations demonstrate how conscious gateway architecture can democratize sophisticated AI capabilities while preserving user agency. Local processing, universal access, and contextual stewardship aren't technical ideals—they're strategic choices that determine whether mobile AI empowers individual capability across diverse contexts and circumstances.

But the same intimate access and ubiquitous reach that could democratize AI capabilities can just as easily enable the most sophisticated behavioral surveillance system in history when different architectural priorities prevail.

Meta AI's integration across Facebook, Instagram, WhatsApp, and Messenger transforms conversational intent into advertising intelligence through a centralized processing system. Although conversations occur within mobile apps, data processing occurs on Meta's servers, with results linked to users' advertising profiles. In October 2025, Meta announced that AI chat interactions would begin serving personalized advertising in December, with no option for users to opt out. This system converts users' most intimate signals—such as intent expressed through natural conversation—into commercial targeting resources.

Life360 and Arity/Allstate Drivewise demonstrate how mobile AI can enable behavioral surveillance while highlighting the benefits of local data processing. Life360, a family safety app, uses on-device machine learning to track location. However, an investigation by The Markup revealed that the company sold precise location data to commercial brokers, despite its safety-focused marketing. Similarly, Allstate Drivewise processes driving events locally but sends behavioral summaries for risk scoring and commercial sharing. This approach shows how mobile AI provides a continuous infrastructure for behavioral extraction while promoting the benefits of local processing for safety. As a result, personal movement patterns are transformed into commodified surveillance data.

Galaxy AI illustrates dependency disguised as device capability. Samsung's S24 series ships with powerful silicon capable of running AI tasks locally. However, compelling features—Live Translate, Chat Assist, generative editing—still require cloud processing through Samsung and Google servers. Enabling "on-device only" processing reduces output quality and disables features altogether. The result is architectural choice rather than hardware limitation: meaningful intelligence remains upstream, and devices function as access terminals rather than self-contained AI agents. Users who prioritize privacy must trade away functionality, revealing how mobile AI can preserve the appearance of local control while reinforcing cloud dependence.

The Architectural Choice Moment

These implementations reveal that mobile AI capabilities exist along spectrums rather than in binary categories, but the architectural choices teams make determine whether sophisticated assistance democratizes opportunity or concentrates control. The promise examples demonstrate engineering decisions that preserve user agency—local processing, model choice, privacy-by-design architecture—while the peril examples show how the same technical capabilities can enable systematic behavioral extraction when different business models and technical priorities prevail.

The window for conscious choice narrows as mobile AI infrastructure becomes entrenched at scale. Infrastructure decisions made now about where intelligence resides and who controls model access will persist for decades, shaping current architectural choices and determining the pace of global AI democratization.

Understanding how strategic leaders, product teams, and policymakers can apply measurement criteria to distinguish democratizing mobile AI from extraction-optimized systems requires practical evaluation frameworks that withstand market pressure.

Evaluating Mobile AI Gateways

Mobile as AI Gateway becomes actionable when organizations can apply consistent measurement criteria to distinguish between systems that leverage mobile's democratizing potential and those that exploit intimate access for corporate extraction. The framework offers practical questions that strategic leaders, product teams, and policymakers can use to evaluate whether mobile AI implementations serve as genuine gateways to user-controlled intelligence.

For Strategic Leaders

Strategic decision makers can evaluate mobile AI investments by examining how proposed systems balance accessibility, capability, and user control across diverse organizational and stakeholder contexts.

How does this mobile AI implementation balance sophisticated capabilities with inclusive access across diverse devices, network conditions, and economic circumstances within our stakeholder ecosystem? Practical implementations are designed to ensure core functionality works across different technical environments while providing enhanced features that do not create significant access barriers. This approach acknowledges that sustainable adoption requires functionality that caters to users with varying resource levels, rather than reserving advanced capabilities solely for premium segments.

What level of transparency and user control does this system provide over data-processing decisions, and how does it accommodate different comfort levels and use cases across our organization? Strategic implementations provide granular control over local and cloud processing, offering clear insight into trade-offs. This allows teams and stakeholders to adjust their engagement based on sensitivity requirements instead of enforcing uniform approaches across different operational contexts.

How effectively does this technology enhance decision-making capabilities through contextual intelligence without creating dependencies that could compromise our operational resilience or stakeholder agency? Organizations that develop sustainable mobile AI capabilities ensure that contextual assistance enhances, rather than replaces, institutional intelligence, allowing teams to remain effective even when AI systems are unavailable and to use advanced support when appropriate.

For Product Teams

Product developers can apply mobile AI gateway evaluation criteria during architecture design and testing phases, recognizing that effective gateway implementations must balance multiple competing priorities while preserving user agency.

How well does our mobile AI architecture perform meaningful functions through local processing while gracefully degrading when cloud connectivity is limited, and what trade-offs does this require in our feature set? Effective gateway design focuses on essential offline functionality while using network access to enhance, rather than enable, core capabilities. This requires conscious choices about which features warrant cloud dependencies and which should be maintained locally.

What range of device capabilities and economic circumstances can access our core mobile AI functionality, and how do our premium features enhance rather than gatekeep essential assistance? True gateway architecture accounts for diverse user contexts and designs core functionality for broad accessibility, ensuring that advanced features enhance capability rather than create fundamental barriers to access across diverse technical and economic environments.

How does our system enable users to understand, control, and modify AI behavior according to their preferences and contexts, while maintaining usability for users who prefer simplified interfaces? Systems designed for genuine empowerment offer transparency and customization without overwhelming users who prefer default settings. This requires designing interfaces that accommodate varying levels of technical sophistication, ensuring that meaningful choices are available to all users.

For Policymakers

Policy frameworks can assess whether they foster democratizing mobile AI architectures while addressing legitimate concerns about privacy, competition, and social impact.

How do our requirements balance user privacy protection with innovation in beneficial contextual assistance, and what implementation approaches do they incentivize among different types of organizations? Effective policies create privacy protections that enable user-centered contextual AI while accounting for how compliance costs and technical requirements affect organizations of various sizes and business models.

What range of technical architectures and business models do our frameworks accommodate for mobile AI development, and how do they prevent anti-competitive practices while encouraging innovation? Policy designs should not inadvertently favor specific technical approaches or organizational structures, while also establishing safeguards to prevent exploitative practices that concentrate capability or extract value without offering proportional benefits to users.

How effectively do our enforcement mechanisms evaluate mobile AI deployment based on user outcomes and capability enhancement rather than compliance metrics that may not reflect actual user agency or benefit? Frameworks serving the public interest assess success by empowering users, broadening opportunities, and implementing effective enforcement strategies that differentiate between beneficial and exploitative practices across various contexts.

Measurement Criteria

Organizations can measure progress through concrete metrics that reveal whether architectural choices serve democratization or enable systematic extraction.

Personal Intelligence Sophistication: *Do users report that their mobile AI becomes more personally useful over time through local learning of individual patterns, or do they experience generic responses that plateau regardless of interaction history?* Authentic personal intelligence enhances individual capability through adaptive strategies, while cloud-reliant systems keep an algorithmic distance that hinders a deep understanding of user context and objectives.

Universal Access Architecture: *Are sophisticated AI capabilities accessible to users with older devices and limited connectivity, or do meaningful features require premium hardware and unlimited data plans?* Conscious gateway implementations make advanced intelligence accessible across various technical environments, while premium-gated systems limit AI capabilities to users who can afford high-end devices and expensive service tiers.

Contextual Intimacy Stewardship: *Do users gain greater control over their digital environment through AI that understands their location, behavioral, and situational patterns, or do they surrender personal information to systems that exploit intimate access for corporate benefit?* Responsible stewardship turns personal context into user empowerment through transparent processes, while extraction-focused systems transform intimate mobile access into surveillance infrastructure disguised as personalization.

When organizations adopt Mobile as AI Gateway as a standard evaluation criterion—applied consistently from initial system architecture to ongoing feature development—they differentiate between mobile AI systems that embrace the democratizing potential of accessible, intimate technology and those that exploit mobile relationships for systematic extraction. The conscious architectural choices made today regarding mobile AI will determine whether advanced intelligence benefits everyone through implementations that democratize access or whether it becomes concentrated among those who can afford premium devices and service plans.

Democratizing AI for All

That moment when I reached for my wallet and found only lint in my pocket revealed something larger than a simple payment inconvenience.

My phone had quietly become more essential than my wallet because it connected me to capabilities that once required separate physical tools—money, identification, insurance cards, membership access, even emergency contacts. But recognizing this shift in my own life helped me understand what's happening globally as mobile devices become the primary computing platform for billions of people.

The architectural choices being made about mobile AI represent something unprecedented in technological history: decisions about whether the most intimate, contextual, and ubiquitous technology humans carry will democratize access to sophisticated intelligence or concentrate it behind premium barriers. While previous computing platforms primarily served professional or recreational functions, mobile devices do so and mediate essential daily activities—communication, navigation, commerce, and healthcare access—for over half the world's population.

The AI capabilities being built into these devices will determine whether artificial intelligence serves human empowerment or corporate extraction for the majority of people who will never own traditional computers.

Mobile as AI Gateway intersects with every other pillar because mobile devices provide the contextual awareness, continuous access, and universal reach that enable the other pillars. Trust-Centered Design becomes essential when mobile AI systems continuously observe behavioral patterns. Human Connection Enhancement depends on mobile AI that reduces rather than increases digital friction in people's daily lives.

Seamless Integration requires mobile devices to serve as the orchestrating platform, coordinating AI capabilities across different contexts and applications. Anticipatory AI relies on mobile devices' unique access to location, schedule, and situational context to understand user intent without invasive surveillance.

What makes this moment in technology both crucial and delicate is the timing. Unlike previous transitions in computing that unfolded over decades, mobile AI capabilities are being developed while the underlying infrastructure remains malleable. This makes current decisions pivotal for global access patterns. Companies that decide on edge computing, cloud dependencies, premium feature gating, and the handling of contextual data are creating patterns that could last for decades, impacting device replacement cycles and platform lock-in.

The window for conscious choice narrows with each billion-dollar infrastructure investment in centralized processing, each premium device launch that gates AI capabilities behind expensive hardware, each privacy policy that normalizes the extraction of intimate data for corporate benefit. Mobile AI architectures deployed at scale create expectations about what's normal, possible, and acceptable in human-AI relationships.

Yet these same architectural choices also shape computational resource consumption and device lifecycle patterns that can either support or undermine the ecological foundations of technological progress.

Chapter 10 -

Pillar 6: Environmentally Responsible Innovation

The Price Being Paid

My friend in the electrical construction industry has been working on data centers in Hillsboro, Oregon, for years. The more I learned from him about these facilities, the starker the reality became: We're building the future faster than we can power it.

The scale of what he described was staggering. Quality Technology Services (QTS) operates massive facilities there spanning 92 acres and 1.5 million square feet. Meta secured one of the largest data center leases in industry history at the campus, requiring 250 megawatts of renewable energy capacity—enough to power 200,000 homes for a year.

What my friend described revealed something profound about how tech companies approach infrastructure development. These facilities can be constructed within two years and are expected to be integrated into existing grids—many already strained and outdated—to absorb massive energy demands that could overwhelm entire communities. A single 30-megawatt facility consumes as much electricity as a whole city. But unlike cities, which develop over decades and fund their own infrastructure, these data centers are often built with the expectation that communities will simply adapt.

Oregon, however, chose a different path.

Instead of allowing tech companies to externalize infrastructure costs to residential ratepayers, lawmakers passed the **POWER Act—"Protecting Oregonians With Energy Responsibility."** The legislation established that large energy users must account for the actual cost of the grid infrastructure they demand, rather than subsidizing it through higher rates for families, many of whom are struggling with high energy bills.

This wasn't an anti-technology policy. What I call **Generational Accountability**—the principle that technological choices should preserve rather than consume finite resources required by future generations—ensures that today's computational decisions don't compromise tomorrow's environmental capacity.

The contrast with other regions is more dramatic. While Oregon required companies to sign long-term contracts covering infrastructure investments, many U.S. data centers are locating in underserved rural communities that lack the political capital or regulatory frameworks to demand such accountability. These communities often absorb the environmental and infrastructure costs while receiving minimal economic benefit.

The construction projects in Oregon that my friend worked on now operate under a completely different model. Companies can't simply demand infinite resources from finite grids. They must demonstrate how their technological advancement serves both business objectives and community stewardship. The result is more responsible innovation that respects both societal needs and planetary boundaries.

Environmental responsibility isn't a constraint on AI development. It's the foundation that enables sustainable progress. When companies are held accountable for their actual ecological costs, they can't avoid responsibility by passing environmental burdens to others—sustainable design becomes a requirement, not an option.

Those buildings in Hillsboro are now effectively fully pre-leased, with their entire power package committed to major anchor customers. The POWER Act approach aims to demonstrate that requiring companies to account for infrastructure costs doesn't stifle development but rather ensures responsible innovation. But Oregon's one of many states with municipalities that need answers to a fundamental question: *How to evaluate whether AI systems worldwide uphold the environmental responsibility necessary for continued innovation?*

Without a way to articulate the stakes, companies can pursue technological advancement while externalizing ecological costs to communities least equipped to bear them.

Defining Environmentally Responsible Innovation

Environmentally Responsible Innovation provides evaluation criteria to distinguish between AI systems that enhance human capability within planetary boundaries and those that sacrifice environmental stability for financial gains. This pillar asks a fundamental question: *How do AI-driven systems being deployed preserve the environmental foundation necessary for sustained technological progress and maintain the capacity for future generations to exercise meaningful choices about their computational resources and innovation paths?*

The pillar measures Generational Accountability along three critical dimensions that teams can apply immediately to their AI implementations:

Conscious Resource Architecture. Current AI systems rely on environmentally extractive infrastructure instead of preserving user choice over computational resources. This dimension assesses whether deployments allow control over the computational footprint and utilize existing infrastructure rather than necessitating new environmental investments.

Stewardship vs. Consumption. AI development can prioritize efficient techniques or rely on resource-intensive architectures that assume unlimited computational resources. This dimension evaluates whether systems become more resource-efficient as they scale and whether they address genuine problems that justify their ecological impact.

Environmental Transparency. Current systems obscure rather than acknowledge the environmental costs of computational choices. This dimension assesses whether stakeholders can access meaningful information about resource usage and whether ecological impacts are noticeable at decision points that could influence behavior.

The urgency arises from the clash between exponential growth and limited resources. If current architectural trends continue, AI systems could consume 15% or more of global electricity within the next decade. However, this situation presents an opportunity rather than a limitation. By designing within planetary boundaries, engineers can discover architectural strategies that not only reduce waste and latency but also enhance privacy and preserve human agency.

Creating environmentally responsible AI systems demands innovation similar to that which transformed room-sized computers into pocket-sized devices—intentional engineering that can produce systems that are more responsive, efficient, and aligned with human objectives than current resource-intensive methods.

Strategic leaders can use this pillar to assess whether AI investments maintain computational choice and transparency for stakeholders, or create environmental dependencies that impose costs on future generations. **Product teams** should evaluate whether their systems use existing infrastructure efficiently while also allowing users to understand and manage their environmental impact. **Policymakers** can examine whether AI frameworks encourage environmental transparency and promote distributed efficiency, or if regulatory gaps permit companies to pursue unsustainable growth patterns that undermine long-term technological capabilities.

This evaluation framework is crucial as we near a potentially irreversible technological threshold. Environmental responsibility is not just one of many factors to consider; it is the foundational reality that must guide all technological advancements. The choices made today by organizations, policymakers, and users regarding environmental accountability will determine whether artificial intelligence can enhance human capabilities sustainably or deplete the planetary resources essential to continued innovation.

Conscious Responsibility

Venho.AI exemplifies Environmentally Responsible Innovation by making architectural decisions that prioritize computational efficiency as a fundamental responsibility rather than merely an external cost constraint. When Antti Saarnio designed Venho's processing approach, he confronted essential questions about the appropriate locations for computation and who should bear the environmental burden. Should AI rely on massive data centers that consume unprecedented amounts of electricity, or could intelligent architecture distribute processing to leverage existing resources more efficiently?

The contrast with the current trajectory of the industry becomes immediately clear in Saarnio's assessment of data center ambitions: "I don't believe that the role the hyperscalers are aiming for is really feasible. It's not sustainable." This skepticism is supported by evidence. Microsoft's CEO recently acknowledged having unused GPUs because the company cannot secure enough electricity to power them. In Finland, the southern regions have reached their electricity grid capacity limits, meaning no additional power infrastructure can be added to support further expansion of data centers.

Instead of viewing rising energy consumption as inevitable, Venho's architecture embraces a core principle identified by Saarnio: "Nature works in a way that it's basically the minimum energy principle. And I think the economy is working in the same principle way as well." This perspective goes beyond environmental philosophy; it highlights that sustainable practices will eventually become economically essential. The current "supercomputer phase" of AI development, as Saarnio describes it, mirrors the 1960s IBM moment when massive centralized computing seemed like the only path forward, until distributed architectures proved more efficient and accessible.

Venmo's technical implementation showcases multiple layers of intelligent resource usage. Instead of processing every incoming email, the operating system allows users to subscribe to specific senders via the front end. As Saarnio explains, "When I get an email from you, I say, okay, let's subscribe to this person's email, and then it's in my memory system." This preprocessing method significantly reduces computational load by focusing solely on information users have explicitly marked as valuable, rather than analyzing every incoming message with AI.

This selective processing allows for the use of smaller models, which further reduce energy consumption. "I can use fairly small models also in the background processing," Saarnio explains, as preprocessing does not require real-time speed. The architecture is designed to prioritize efficiency over immediacy for tasks that do not need instant responses. Venho's 3B parameter model is not only more cost-effective than incurring cloud usage fees but also significantly less energy-intensive than larger models that rely on data center infrastructure.

The approach also leverages computational resources that users already possess. When advising industrial clients, Saarnio asks a question that challenges the assumptions in conventional AI deployment: "Why would you acquire huge computing capacity if you have desktops already, which can process a third of the computing for the AI? And you could have only a third for the cloud, which you use when you need it, basically."

This distributed processing philosophy recognizes that organizations have already invested in desktop computing infrastructure. Rather than leaving that capacity idle while purchasing cloud processing, Venho's node architecture enables existing hardware to handle local AI tasks efficiently. Users download Venho on their desktops, and those machines become active participants in a distributed AI ecosystem rather than passive terminals dependent on remote servers.

"Our principle is that let's use the computing capacity that people already have," Saarnio explains. The Jolla Mind2 device serves as an optional server node for users who want dedicated local processing power. But the architecture doesn't require specialized hardware. It works with the computers and devices users already own. This approach fundamentally challenges what Saarnio calls "the craziness" of big players trying to bind users to centralized infrastructure for economic gain.

The environmental impacts go beyond just individual energy savings—they also affect overall resource efficiency. When AI processing is distributed across existing devices instead of being centralized in large data centers, the load on the infrastructure is reduced. This approach uses hardware already manufactured and deployed, reducing dependence on freshwater-based cooling systems. As a result, fewer power plants will need to be built to meet computing demands, and there will be no need for specialized cooling systems that release concentrated heat into local ecosystems.

The architectural choices underlying Venho's environmental approach reveal how resource responsibility can serve the trinity of forces that define the Mobile Era of Intent. **Agency** remains intact because users control their computational resources and choose when to leverage cloud processing versus local capability. **Trust** develops through transparent resource usage that doesn't hide ecological impact behind abstracted cloud services. **Intent** becomes the organizing principle that determines which computations justify the use of energy.

This represents the fundamental choice between AI development that prioritizes economic gain at the expense of the environment and an architecture that treats sustainability as a design principle worth engineering. When computational efficiency is treated as a necessity rather than an afterthought, the resulting systems emerge from engineering strategies that wouldn't have been developed under the assumptions of unlimited resources.

Responsible Innovation In Practice

Environmentally Responsible Innovation becomes distinguishable when we examine how different organizations use the same AI capabilities through different architectural choices. The computational power used to manage renewable energy systems or enhance resource efficiency can also be directed toward accelerating fossil fuel extraction or creating complex infrastructure that consumes significant resources.

Recent deployments reveal dramatic contrasts between systems designed to expand future capacity through conscious resource stewardship and those that externalize resource costs while optimizing for corporate convenience. When teams face decisions about energy architecture, measurement frameworks, and business model alignment, their choices either honor generational accountability or accumulate technological debt passed on to future generations.

Puget Sound Energy's (PSE) virtual power plant, powered by AutoGrid Systems, demonstrates how AI can orchestrate existing renewable resources rather than requiring new infrastructure. The system, essentially like a "crowd-sourced" power grid, coordinates tens of megawatts across participating customers' thermostats, batteries, and EV chargers, enabling PSE to balance renewables and reduce fossil backup generation across its 1.2 million-customer service territory. By intelligently coordinating micro-decisions like pre-cooling homes by one degree or delaying EV charging to off-peak hours, PSE's AI system eliminates the need for new peaker plants through distributed demand management.

CodeCarbon's open-source library demonstrates transparency as infrastructure by enabling any developer building AI-driven experiences to track energy consumption and estimated CO2 emissions directly within their code, logging environmental impact alongside performance metrics in real-time dashboards. Whether training models, running inference in production apps, or deploying recommendation systems, teams can measure actual computational carbon costs rather than relying on provider sustainability claims. When product teams discover that one AI feature configuration significantly reduces emissions while maintaining an equivalent user experience, environmental responsibility becomes a standard engineering decision, creating systematic pressure for efficiency across organizations deploying AI capabilities.

Syntiant's Neural Decision Processors (NDPs) demonstrate that AI architecture can prioritize energy efficiency by design rather than by mitigating consumption afterward. Their line of deep learning edge processors runs voice and sensor recognition models locally on tiny, battery-powered devices. The company's TinyML development board demonstrates this approach, consuming just 140 microwatts—roughly the power of a fingernail-sized solar cell in sunlight. In industry-standard efficiency testing across multiple AI tasks, Syntiant's Core 2 architecture achieved the lowest energy consumption per inference, proving that practical intelligence doesn't require megawatt data centers. This approach prioritizes energy efficiency in system design alongside accuracy and latency.

These implementations demonstrate conscious resource architecture in practice: using Environmentally Responsible Innovation to expand what existing infrastructure can accomplish rather than accelerating resource consumption to feed large algorithmic appetites.

But the same AI capabilities that enable environmental stewardship can just as easily perfect extraction when different architectural choices prevail.

AI-optimized fossil fuel extraction systems pose a serious risk of depleting reserves that climate scientists warn must remain untapped. Global Witness, an independent investigative and campaigning organization, reports that major oil and gas companies are using AI to speed up the extraction of new reserves. These systems enhance drilling paths and lower exploration costs. Although these advancements are presented as "efficiency improvements," they compromise generational accountability by treating technological capabilities as a justification for expanding operations. Climate scientists caution that such expansions will lead to emissions exceeding safe levels.

AI-optimized cryptocurrency mining highlights how improvements in individual efficiency can obscure the overall environmental impact. Research conducted by Carnegie Mellon University and the Open Energy Outlook (OEO) Initiative reveals that while crypto mining operations use AI to enhance their individual efficiency, their collective expansion necessitates costly upgrades to transmission infrastructure and backup power systems. This, in turn, raises electricity costs and increases emissions for all grid users. When AI optimization allows individual operators to seem more responsible, but the industry's overall growth forces communities to bear the burden of infrastructure upgrades, it leads to environmental costs being disguised as technological progress.

AI training water consumption highlights how claims of computational efficiency can obscure significant resource extraction at the infrastructure level. While tech companies market AI models as "cloud-based" solutions that eliminate the need for local resources, the physical infrastructure supporting these systems actually consumes millions of gallons of water each day to cool data centers during intensive training sessions. Companies generally report total water usage for data centers without specifying consumption related to particular AI workloads. This lack of detailed information makes it impossible for customers or policymakers to assess the actual environmental impact of deploying large language models compared to more efficient alternatives. Consequently, this allows companies to shift the costs onto others, reducing their ecological accountability.

These examples illustrate how environmental responsibility presents challenges in measurement and transparency. When AI systems prioritize ecological impact over short-term profits, advanced technology can eliminate rather than shift environmental costs.

The Choice Between Stewardship and Extraction

The stark differences between these implementations show that environmental outcomes depend heavily on the evaluation frameworks teams use during AI development. When organizations prioritize the design for energy and resource consumption, remarkable innovations can arise. These include virtual power plants that manage renewable energy, transparency tools that make carbon impacts visible, and edge processors that demonstrate that intelligence can require only microwatts instead of megawatts. These capabilities can either promote generational accountability or hasten the transfer of ecological debt.

A concerning pattern emerges when companies use AI to accelerate resource extraction while shifting environmental costs to future generations through complex technologies. This includes practices such as accelerating fossil fuel extraction, optimizing energy-intensive operations that overload shared infrastructure, and concealing significant resource consumption behind "cloud-based" solutions. These actions highlight a fundamental disconnect between technological progress and care for the planet—leveraging intelligence to enhance resource extraction instead of fostering thoughtful decisions about resource use.

Organizations that consistently implement criteria for evaluating Environmentally Responsible Innovation gain a competitive edge by genuinely improving resource efficiency rather than resorting to complex cost-cutting measures that are presented as innovation. However, achieving this goal involves more than conscious choices by individual companies—it requires an evaluative framework that ensures that environmental transparency is a technical necessity, not merely an ideal to strive for.

Evaluating Environment-Honoring Innovation

Environmentally Responsible Innovation becomes actionable when organizations use consistent measurement criteria for their AI applications. This framework offers practical questions to help differentiate between systems that support environmental sustainability for long-term technological advancement and those that exploit ecological resources for short-term corporate gain.

For Strategic Leaders

Strategic decision makers can evaluate AI investments by examining whether proposed systems will operate within environmental boundaries that enable long-term technological capability or create dependencies on resource consumption patterns that could undermine future innovation.

Does this AI implementation employ efficiency techniques that reduce resource dependency, and does it become more efficient as it scales, rather than consuming additional resources? Environmental stewardship approaches require frameworks that assess whether AI capabilities leverage technical approaches such as edge processing, model optimization, and distributed intelligence to deliver performance within planetary boundaries, and whether systems address authentic problems that justify their ecological footprint. Systems that employ quantization, pruning, or compression techniques enable local processing to reduce dependence on data centers and to become more resource-efficient as they scale.

To what degree does this AI system enable our organization to choose where and how processing occurs, and what control do we maintain over our computational footprint as environmental constraints evolve? Conscious resource architecture requires frameworks that assess whether AI deployments maintain an organization's control over computational resources or create dependencies that limit their processing options. Effective systems should enable choices between cloud and edge deployments, allow adjustments to processing intensity based on resource availability, and distribute intelligence to optimize the use of existing infrastructure capacity.

What visibility do we have into this AI system's actual resource usage, and can we track and respond to resource consumption patterns over time? Environmental transparency requires frameworks that assess organizations' access to meaningful information on computational resource usage, the visibility of environmental costs in deployment decisions, and the verification of sustainability claims against consumption data. Transparent reporting on energy use and resource efficiency supports conscious ecological choices, while cloud abstractions that obscure consumption hinder organizations' ability to manage their environmental impact.

For Product Teams

Product developers can apply Environmentally Responsible Innovation criteria during architecture and design phases to ensure AI features enhance human capability within resource constraints rather than optimizing for metrics that externalize environmental costs.

How effectively does our AI architecture balance performance capabilities with resource efficiency, and which design choices enable users to calibrate this balance to their environmental priorities and computational constraints? Technical implementations should provide users with meaningful control over processing intensity, allowing them to make conscious trade-offs between AI sophistication and resource consumption. Options for edge processing, model selection, and efficiency settings can indicate whether systems prioritize user agency over default settings that are resource-intensive and assume unlimited environmental capacity.

What transparency does our system provide about the environmental impact of different AI features, and how do we enable users to make informed decisions about which capabilities justify their resource requirements? Ecological responsibility requires interface design that makes resource consumption visible rather than abstract. Systems that provide energy impact indicators, transparency into processing location, and feature-specific ecological costs enable conscious user choice about their level of engagement with AI, instead of necessitating passive acceptance of hidden resource consumption.

How does our AI system's resource-consumption pattern change as it scales across different user populations and usage contexts, and what architectural decisions ensure efficiency gains rather than linear resource scaling? Product teams should evaluate whether their AI systems become more environmentally efficient as adoption grows through shared processing, model optimization, and distributed intelligence. If the systems require proportional infrastructure scaling for each additional user, it may suggest that architectural decisions prioritize vendor expansion over environmental optimization.

For Policymakers

Policy frameworks can encourage environmentally responsible AI development by creating transparency requirements and accountability mechanisms that align technological advancement with ecological stewardship.

Do our regulatory approaches require meaningful transparency into the environmental costs of AI systems, enabling organizations and individuals to make informed decisions about their computational consumption patterns? Policy frameworks should require the disclosure of ecological impacts associated with AI services, including energy consumption, carbon footprint, and resource utilization metrics. Effective regulations empower individuals and organizations to make informed decisions about engaging with AI, rather than compelling them to accept hidden environmental costs that are subsidized by public infrastructure and imposed on future generations.

How effectively do our standards encourage systems that distribute intelligence efficiently across existing infrastructure rather than concentrating processing in energy-intensive centralized facilities that strain shared resources? Regulatory approaches should promote edge computing, distributed processing, and the integration of renewable energy while preventing the avoidance of environmental costs through cloud abstraction. Policies that reward genuine efficiency improvements, rather than resource concentration, help distinguish genuine environmental stewardship from more complex forms of cost-shifting.

What mechanisms do we have for ensuring that AI development serves genuine human needs and societal benefit that justify environmental resource allocation, rather than optimizing for engagement metrics that may conflict with public well-being? Environmentally Responsible Innovation requires policy frameworks that assess whether AI capabilities enhance human and societal capacities while justifying their ecological impact. Regulations should distinguish between systems that effectively address real issues and those that offer advanced solutions primarily for corporate profit rather than for the public good.

Measurement Criteria

Organizations can measure progress through concrete metrics that reveal whether architectural choices support resource stewardship or enable passing environmental burdens to future generations.

Conscious Resource Architecture: *Are users and organizations able to make informed decisions about their computational resource consumption, or do they become dependent on systems that obscure environmental costs through cloud abstraction and predetermined processing choices?* Genuine resource consciousness enables effective control over processing location, model complexity, and energy consumption, based on individual priorities and environmental capacity.

Stewardship vs. Consumption: *Do AI deployments become more resource-efficient as they scale and solve authentic problems that justify their environmental footprint, or do they default to resource-intensive architectures that assume unlimited computational capacity?* Environmental stewardship creates systems that improve efficiency over time while providing genuine capability. In contrast, consumption patterns increase resource demands without generating commensurate value.

Environmental Transparency: *Can users understand the ecological impacts of different AI features and make conscious choices about their computational engagement, or are ecological costs obscured by technical complexity, preventing informed decision making?* Transparent environmental accountability offers insight into resource consumption, renewable energy utilization, and efficiency improvements, facilitating informed choices about AI adoption and usage levels.

When organizations make Environmentally Responsible Innovation a standard criterion for evaluation—applied consistently from the initial AI architecture to ongoing feature development—they can develop capabilities that promote Generational Accountability. These organizations develop AI systems that address real-world problems while adhering to environmental guidelines. They achieve efficiency gains that accumulate over time and help preserve environmental resources for future technological innovation. This approach ensures that today's computational decisions enable future generations to make meaningful choices about their environment rather than compromising those choices.

The Environmental Imperative

What my friend shared with me about the Hillsboro data centers revealed more than construction projects or energy policy. Those facilities represent a typical pattern playing out globally: sophisticated technology designed to extract value while externalizing costs to communities that lack the political power to demand accountability. Oregon's POWER Act wasn't just energy regulation—it was a rare example of policy courage that refused to subsidize corporate expansion with community resources.

Oregon isn't alone in this approach, as more communities and municipalities are demanding societal accountability from Big Tech's expansion. These lawmakers chose conscious responsibility over the path of least resistance, forcing companies to pay the actual infrastructure costs their demand creates rather than shifting that burden to residential ratepayers already struggling with rising costs.

Yet policy courage alone can't solve the systemic challenge. This is why Environmentally Responsible Innovation isn't just another evaluation criterion—it's the foundation that determines whether the Mobile Era of Intent serves human agency or corporate gain. The computational arms race, intensifying global temperature pressures toward the 1.5°C threshold, shows what happens when technology optimizes for corporate benefit while ignoring environmental impact.

Establishing this environmental foundation requires the other pillars to work together. Human Connection Enhancement ensures AI reduces digital noise rather than creating resource-hungry engagement systems. Trust-Centered Design provides transparency into computational costs instead of hiding environmental impact behind cloud abstractions.

Seamless Integration coordinates efficiently across platforms rather than forcing users to run multiple energy-intensive systems. Anticipatory AI understands context once, rather than constantly reprocessing user intentions. Mobile as AI Gateway distributes intelligence to edge devices rather than concentrating everything in massive data centers.

What makes this moment historically unique is that artificial intelligence serves as both an accelerant and an antidote. The same computational capabilities that can enhance cost externalization also provide the transparency, distributed efficiency, and resource optimization needed for environmental stewardship. Every decision made—whether in product meetings, infrastructure planning, or policy frameworks—represents a choice about how to direct these capabilities.

The environmental imperative indicates that the Mobile Era of Intent can thrive only if AI effectively serves genuine human needs while respecting planetary boundaries. This means that intelligence and efficiency must work together rather than compete. The architectural choices organizations make cannot be postponed for future updates or improved technology—they must be integrated into every system, business model, and policy framework established today.

A truly sustainable future will depend on the ability to consciously choose AI that aligns with human intent and operates within environmental limits that enable technological advancement.

Chapter 11 -

The Path Forward

Maintaining My Motorcycle

I love working on my dirt bike. There's something satisfying about the mechanical precision, patience, and problem solving involved in this process. Everything has a specific sequence and a proper way to fit together, making it a meditative experience—what I like to call "two-wheel therapy." This work demands my full attention, free from distractions, and allows me to develop my skills. When I can fix something myself, I get back on the trails faster and gain a better understanding of my bike.

But I learned that having proper guidance without the right tools will get me nowhere, just as having the right tools without clear guidance often leads to costly mistakes.

The clutch replacement should have been straightforward. The YouTube guide was exceptional—clear visuals, step-by-step explanation, and all the torque specifications. I had my workspace organized, tools laid out, parts ready to go. Everything made perfect sense until I reached one crucial step: removing the clutch hub. The guide showed exactly what needed to be done, but I didn't have the specialized pliers required. No amount of understanding could substitute for the missing tool.

Instead of riding that weekend with my friends, I ended up at the motorcycle shop, paying for a service I could have performed myself if I'd had the right equipment.

A few weeks later, I decided to upgrade my headlight. This time, I made sure I had all the tools I needed—the LED light assembly, a heated soldering iron ready, shrink tubing cut to size, and wire strippers on hand. I was fully prepared for precise electrical work. But the wiring diagram in the instructions was incomplete. It showed which wires to connect, but not which specific colors went where on my model year. I had the capability but not the clarity.

I could have fumbled around testing the wires one by one, but instead I found myself again at the shop, frustrated that I'd come so close to completing the work myself.

Both times, the solution was obvious in retrospect. I needed the knowledge and the tools to work together. When I had clear instructions but lacked the equipment, knowledge alone couldn't get the job done. When I had the proper tools but incomplete instructions, skill without guidance led nowhere.

The work got done eventually, but only when both elements aligned—when someone with both the knowledge and the tools could complete what I'd started.

I realized this pattern extends far beyond motorcycle maintenance. In any complex endeavor, there is a need for both a clear understanding of what needs to happen and the right resources to make it happen—one without the other results in costly delays and unnecessary frustration.

We stand at a similar moment with the current AI trajectory. The technological tools now exist to build AI systems that understand and serve human intent—teams like Antti Saarnio's at Venho.AI have demonstrated this in practice. There now exists a framework—the Six Pillars of Intent—that offers clear guidance for distinguishing AI that enhances human capability from AI that exploits human psychology.

The real question isn't whether technology can be developed to address fundamental human needs rather than just serve corporate interests. Instead, the key question is whether teams will use both the tools and the framework together, or default to inevitability by deploying one without the other. For the first time, we have both advanced AI capabilities and evaluation criteria to distinguish between implementations that benefit humans and those that do not.

The window for conscious choice is narrow, but the path is clear. When organizations integrate the Six Pillars with the AI capabilities being deployed today, we can finally align the most powerful technology of our era with genuine human intent.

The Framework for Intent-Driven AI

The Six Pillars of Intent are most effective when viewed not as individual evaluation criteria, but as an integrated decision-making framework that operationalizes the trinity of forces shaping intent-driven AI in development. Just as motorcycle repair requires both guidance and tools to work together, making conscious choices about AI requires a solid philosophical foundation paired with a consistently applied practical methodology.

The Six Pillars are not a checklist—they form a comprehensive measurement system that assesses whether technology choices support or undermine human agency, respect or exploit trust investments, and align with genuine intent versus corporate convenience. When used as a unified framework, they reshape how organizations approach technological progress.

The Trinity of Forces in Practice

These three forces, when put into practice through the Six Pillars, determine whether AI development enhances human capability or optimizes for institutional extraction.

Agency emerges through specific architectural choices that preserve human understanding and control over AI behavior. Agency isn't an abstract choice—it's whether users can comprehend how systems learn about them, contest decisions that don't serve their goals, and exit relationships without losing their digital autonomy. The Six Pillars reveal agency operating through transparency mechanisms that enable meaningful recourse, business models that align with user interests rather than exploit them, and cognitive coordination that enhances rather than replaces human capability.

Trust is an active relationship that requires systematic stewardship through conscious design choices. Trust isn't a binary resource to be honored or exploited—it's a conditional relationship continuously shaped by whether AI systems protect vulnerabilities or exploit them, whether anticipatory capabilities serve user goals or manipulate behavior, and whether processing architectures preserve user sovereignty or concentrate control. Trust builds through verified transparency, genuine user empowerment, and alignment between stated purposes and actual system behavior.

Intent functions as the organizing principle that enables AI to understand and serve human goals rather than requiring humans to adapt to system logic. Intent isn't just what we want technology to accomplish—it's the bridge between human thinking patterns and technological capability that preserves human authority over outcomes. Intent-driven systems predict needs to help users achieve goals rather than predicting behavior to influence decisions, creating technology that adapts to human cognitive patterns while maintaining human agency over complex processes.

The Pillar Implementation Methodology

Each pillar functions as an evaluation lens, revealing how design choices either strengthen or undermine the trinity of forces. Applied systematically, they transform abstract principles into measurable criteria that distinguish human-serving implementations from extractive ones.

Human Connection Enhancement assesses whether attention and relationship technologies preserve user agency in AI learning, foster trust by protecting psychological vulnerabilities rather than exploiting them, and promote genuine connection intent instead of merely optimizing for engagement.

Trust-Centered Design evaluates if system architectures allow meaningful user control, establish conditional trust relationships through aligned incentives and transparent operations, and prioritize user goals over corporate data collection.

Seamless Integration evaluates whether coordination technologies preserve cognitive authority over complex processes, maintain trust by enhancing rather than replacing human capabilities, and serve as a direct expression of intent instead of forcing adaptation to system logic.

Anticipatory AI assesses whether its predictive capabilities honor human decision-making authority, foster trust through transparent reasoning that users can verify and correct, and prioritize understanding of goals over manipulating behavior.

Mobile as AI Gateway assesses whether mobile AI architectures maintain user control over personal intelligence, respect the trust placed in intimate device access, and prioritize human cognitive adaptation over corporate surveillance interests.

Environmentally Responsible Innovation evaluates if computational choices promote conscious resource management, foster trust through environmental transparency, and prioritize genuine efficiency over convenience in extractive business models.

The trinity evaluation becomes systematic when pillars work together—transparency enables agency, aligned incentives build trust, and shared understanding of goals serves authentic intent across all implementations.

A Proof of Concept

Venho.AI demonstrates what becomes possible when teams possess both technical capability and an intuitive understanding of human-centered principles. Antti Saarnio's team achieved remarkable integration—edge processing that preserves trust while reducing environmental impact, semantic interaction that eliminates silos while enhancing human connection, and user-controlled memory that enables anticipatory assistance without sacrificing agency. Every architectural choice reinforces the others.

But Saarnio's team accomplished this through exceptional leadership intuition and mission-first thinking. Imagine what becomes possible when teams start with both technical capabilities and explicit methodology from day one; when development teams have not only the tools to build human-centered AI but also clear guidance for navigating the complex trade-offs involved in implementation; when product managers can apply systematic evaluation criteria rather than relying on individual insight to distinguish enhancement from extraction.

Organizations equipped with both technological capability and conscious choice frameworks don't just build successful products—they systematically transform how technology development happens. They establish a sustainable competitive advantage through human-centered innovation rather than a race to the bottom that ignores environmental costs. They provide evidence that shifts industry standards toward conscious implementation.

The framework stops being about evaluation and starts being about transformation. When teams recognize that agency, trust, and intent operate as an integrated system, and when they apply a pillar-based methodology during moments when technology direction is determined, conscious choice becomes practical rather than aspirational. The potential is extraordinary: not just individual success stories, but systematic redirection of technological momentum toward human flourishing.

From Evaluation to Implementation

The implementation insight is straightforward: the Six Pillars of Intent become more than evaluation criteria when teams consistently apply them at the moments when a technology's direction is determined. Like having both the right tools and clear instructions, the framework transforms from an assessment checklist into a decision-making methodology—but only when applied at choice points that actually shape how AI gets built and deployed.

This transformation happens through consistent application across four types of decisions that collectively determine whether AI development serves human intent or something else entirely. The pillars work best when they guide the choices being made right now, in product meetings, vendor evaluations, and design decisions, rather than as post-development audits of systems already built.

Redirecting Organizational Momentum

Strategic leaders shape AI implementation by setting investment priorities, defining success metrics, shaping competitive positioning, and establishing organizational policies. The Six Pillars offer practical guidance for decision makers who want their AI investments to enhance, rather than erode, human capability and stakeholder relationships.

Investment Decisions require applying Trust-Centered Design criteria to vendor selection and internal development priorities. Instead of choosing AI systems solely on the basis of efficiency metrics or cost reduction, leaders can evaluate whether proposed implementations preserve meaningful human control, provide transparency into decision-making processes, and align business-model incentives with user interests. This means asking vendors to demonstrate how their systems enable user agency rather than just showcasing performance benchmarks.

Success Metrics shift from pure efficiency optimization toward Human Connection Enhancement measurements alongside traditional productivity gains. Organizations can track whether AI implementations create more time for high-value human interaction, reduce cognitive overhead that fragments attention, or enable teams to focus on creative collaboration rather than routine coordination. This doesn't eliminate efficiency metrics—it balances them with measurements that reveal whether technological progress addresses human needs.

Competitive Positioning uses the framework to differentiate from competitors who optimize for engagement extraction rather than capability enhancement. When organizations can articulate why their AI approach preserves user agency, honors trust investments, and serves genuine human intent, they build competitive advantage through conscious choice rather than race-to-the-bottom optimization. Customers increasingly recognize the difference between AI that serves them and AI that serves advertising metrics.

Policy Architecture embeds conscious choice criteria into procurement processes, partnership agreements, and development guidelines. This means establishing organizational standards that require Anticipatory AI capabilities to enhance, rather than replace, human decision making; Seamless Integration approaches that reduce, rather than increase, cognitive burden; and Mobile as AI Gateway implementations that enhance personal intelligence capabilities while protecting the contextual intimacy that makes mobile AI valuable.

Building Intent-Driven Capabilities

Product design and development teams transform evaluation criteria into an implementation methodology by incorporating pillar-based assessment into design processes, architectural decisions, user empowerment features, and responsible resource use.

Design Integration incorporates the Six Pillars evaluation into sprint planning and feature specification. Product teams can apply Human Connection Enhancement criteria when designing social features, ensuring that these features amplify authentic relationships rather than simply optimizing for engagement metrics. This means building features that help users accomplish their goals more efficiently, so they have more time for what matters to them personally, rather than features designed to maximize time on platform.

Architecture Decisions employ the principles of Seamless Integration and Mobile as an AI Gateway to inform technical implementation decisions. Teams can design systems that intelligently orchestrate user workflows across existing platforms, rather than creating new silos that require additional coordination. This approach involves building AI capabilities that understand user context and adapt to individual needs, rather than requiring users to rely on additional technological dependencies.

User Empowerment focuses on applying Anticipatory AI principles to enhance human decision making instead of creating reliance on algorithms. Development teams can create systems that present relevant information and coordinate resources while maintaining user control over their choices. This means designing AI that offers options and suggests actions that users can accept, modify, or ignore, rather than implementing systems that make decisions without significant human oversight.

Environmental Consciousness integrates Environmentally Responsible Innovation into computational and infrastructure decisions. Teams can choose edge computing architectures that reduce energy consumption, implement efficient algorithms to accomplish the same tasks with fewer resources, and design systems that scale sustainably without imposing massive computational overhead on basic functionality.

Creating Systemic Incentives

Policymakers can operationalize conscious AI choice by establishing regulatory frameworks, enforcement mechanisms, market structures, and coordination standards that systematically reward human-serving implementations over extraction-focused optimization.

Regulatory Framework Development applies Trust-Centered Design principles to distinguish AI systems that honor user agency from those that exploit psychological vulnerabilities. Policymakers can establish requirements for meaningful transparency that enable user control, business model accountability that aligns providers' incentives with users' interests, and recourse mechanisms that preserve users' autonomy over algorithmic decisions affecting employment, housing, lending, or health outcomes.

Enforcement Mechanisms use Human Connection Enhancement criteria to protect vulnerable populations from AI systems that exploit emotional attachment or social isolation. Policy frameworks can establish requirements for Trust-Centered Design, including safety guardrails for AI companion systems, mandate mental health oversight for anticipatory intervention technologies, and set liability standards for systems that prioritize engagement optimization over user well-being—particularly when algorithmic manipulation could undermine human agency or cause physical or psychological harm.

Market Structure Reform leverages Mobile as AI Gateway and Environmentally Responsible Innovation principles to prevent extraction-focused business models from dominating AI development. This involves creating competitive advantages for organizations that preserve user sovereignty over personal AI processing, demonstrate genuine environmental stewardship through transparent resource accounting, and enable user choice through interoperability standards rather than platform lock-in mechanisms.

International Standards Coordination applies Seamless Integration frameworks to establish cross-border protocols that preserve user agency while enabling beneficial AI capabilities. Policymakers can develop standards that allow AI systems to understand user context across jurisdictions without creating surveillance infrastructure, facilitate human-centered technology transfer, and coordinate regulatory approaches that reward intent-driven innovation while preventing a regulatory race to the bottom.

The Transformation Pattern

When applied consistently at crucial decision points, the Six Pillars of Intent help organizations shift their momentum toward technologies that align with human intent, maintain valuable agency in the face of inevitability, and respect decades of trust investments. This framework not only helps teams evaluate existing AI but also transforms the decision-making process for new technologies. It establishes clear criteria to differentiate between enhancement and extraction at pivotal moments, guiding implementation choices.

This transformation happens gradually through daily decisions driven by conscious choice principles instead of default optimization. Each decision based on the pillars creates opportunities for the next conscious choice, incrementally shifting the path of least resistance toward AI development that enhances human capabilities.

The tools and guidance are available together for the first time. The question is whether organizations apply them consistently enough to shape technology that promotes human flourishing instead of serving corporate convenience.

Becoming a Conscious Agent

If you believe that the results of technology stem from conscious design choices rather than being the result of inevitable progress—if you believe that the decades of trust invested in these systems deserve to be honored rather than exploited—if you believe that AI can understand and serve human intent rather than merely manipulating human attention—then you already have what is necessary for conscious agency: the recognition that different outcomes are possible.

This isn't about transforming everything at once or carrying the weight of technological history on individual shoulders. Conscious agents consistently apply evaluation frameworks across their sphere of influence—whether it's personal AI tools, team development decisions, organizational technology strategy, or policy frameworks that shape market incentives. The scope matters less than the consistency.

This consistency matters because the Six Pillars of Intent work as evaluation criteria, not pass/fail requirements. A team might excel at Trust-Centered Design even as it develops its Anticipatory AI capabilities. An organization might make significant progress in Human Connection Enhancement while advancing better Environmental Responsibility Innovation.

The framework's power lies in providing language and systematic evaluation for conscious choices, not in demanding perfect implementation across all criteria simultaneously. Conscious agents use the pillars to shape direction and measure progress, recognizing that working within the framework is more important than achieving immediate optimization across all dimensions.

Each conscious choice creates space for others. When a product manager prioritizes authentic relationships alongside performance metrics, they make it easier for designers to focus on meaningful human interaction. When a policymaker establishes transparency and user control requirements, they create a competitive advantage for organizations that honor user agency.

When a strategic leader measures AI deployments by both enhancing capabilities and improving efficiency, they shift organizational momentum toward technologies that serve human intent. Teams equipped with explicit evaluation criteria can advocate for human-centered approaches even under pressure to optimize for corporate metrics. Organizations that demonstrate viable alternatives to extraction-based AI create proof points that influence competitive dynamics.

The Mobile Era of Intent isn't something that will happen to us—it's a new era being created right now through daily decisions guided by explicit principles about what technology should serve. The Six Pillars of Intent provide the evaluation framework. Current technological capabilities provide the tools for implementation. This convergence creates unprecedented opportunity for conscious choice to shape AI development.

But opportunity requires application. The framework doesn't implement itself. The tools don't choose their own direction. The window for conscious choice narrows.

Conscious agents recognize this moment for what it is: the threshold at which technology finally can understand human intent, where evaluation frameworks exist to distinguish enhancement from extraction, and where market forces increasingly favor alternatives to surveillance-based optimization. The elements necessary for conscious choice exist simultaneously.

The choices ahead shape AI development toward human flourishing, direct trust investment toward genuine capability enhancement, and enable technology to adapt more effectively to human needs. Choices guided by conscious agents who refuse to accept that technological progress serves only those who build it, rather than those who live with it.

The future we deserve will not emerge by default. It will be shaped by design, and forged by intent.

About the Author

Van E. Eiseman was thrust into leadership at the age of 18, when the U.S. Navy recognized his potential before he realized it himself. For over 30 years, he has shaped technology strategy, guiding organizations in designing digital experiences while asking harder questions about who they serve.

Then, in 2014, Van was diagnosed with Stage 4 tonsil cancer. Survival required perseverance, patience, and faith in healing that could not be rushed. The experience left him with a lasting belief: that endurance is sustained not by will alone, but by a sense of responsibility beyond oneself.

That belief shapes how he thinks about who gets to decide, who bears the cost of those decisions, and what happens when technology moves faster than our willingness to pause and ask better questions.

Van lives near Portland, Oregon, with his wife, Gena, and their children, Lance and Maya. When he isn't writing or advising, he enjoys playing guitar, riding dirt bikes through the forests of the Pacific Northwest, and hand-painting custom Vans footwear.

Acknowledgements

Gena — You believed in this before I did. You believed in me when I didn't. Your quiet support, your "keep going" when I questioned everything, your eye for what readers need—you've done this before, and it showed. I couldn't have finished without you beside me. TQM!

Lance — You're the skeptic I needed. The videos you shared, the questions you raised about AI and authenticity—they forced me to think harder. This book exists for your generation. I wanted to write something you'd welcome, not dismiss. I love you, buddy!

Maya — Every word from you these nine months has been genuine encouragement. Not blind faith—real confidence. You built me up when I wavered. How could I not keep going? Your strength is always in my pocket, just when I need it. I love you, kiddo!

Deb — Thank you for your guidance, wisdom, and support. When we spoke about this as more than a collection of technologies—that what I was explaining was a societal shift—it set everything in motion. I am truly grateful for your care and so much more. Love you, sis!

Al — Since you came into my life, you've brought clarity and purpose. Not only as a colleague and career partner, but also as a bandmate and best friend. Sitting down with you, exploring trust dynamics with AI, was the lever that moved my work forward. I am eternally grateful. Love you good!

Brynna — Since the days at Fado working on the animation project (I will finish it one day), I've admired your creative warmth and generosity. When I shared the initial manuscript, your encouragement and the gift of promise and peril gave the book clarity. And "let the reader write their own story"—that insight kept me going. Thank you, my friend!

Craig — Thank you for helping me give the pillars strength as guideposts—the navigation, the action to make this more than a collection of articles, but a framework for change. I am beyond grateful for our partnership. What we built stands true / the path ahead is ours now / onward and upward

Antti — Thank you for your time, your validation of the Six Pillars, and your generosity in letting Venho and Mind2 take the spotlight they deserve. I'm excited to see where leaders like you take us. Kippis!

Peter — Your feedback challenged every word, and I wouldn't trade any of it. You taught me to write with better clarity and purpose. Appreciate you!

Jesse — You did more than tighten wording and punctuation—you gave me confidence in every word. I cannot thank you enough.

And to Pepper and Nacho — My constant writing companions, who asked for nothing but presence, a steady stream of treats, and our nightly walks. Woof!

References

The following sources informed the research, examples, and analysis presented throughout this book. These references provide additional context for readers interested in exploring specific topics in greater depth, verification of cited statistics and studies, and pathways for continued learning about AI implementation, technology ethics, and human-centered design principles.

Chapter 1

Agency

Patrick Spencer "The 2025 AI Security Gap: Why 83% of Organizations Are Flying Blind," Kiteworks, August 20, 2025.
https://www.kiteworks.com/cybersecurity-risk-management/ai-security-gap-2025-organizations-flying-blind/

Devansh, "Busting Unions with AI: How Amazon Uses Algorithms Against Workers," Artificial Intelligence Made Simple, May 26, 2025.
https://artificialintelligencemadesimple.substack.com/p/busting-unions-with-ai-how-amazon

Inevitability

Timothy Prestianni, "59 AI Job Statistics: Future of U.S. Jobs," National University, May 30, 2025.
https://www.nu.edu/blog/ai-job-statistics/

LeadInnovationZ, "Consumer Demand for Transparency," LinkedIn, April 28, 2025.
https://www.linkedin.com/pulse/consumer-demand-transparency-leadinnovationz-gw94c/

The Trust Paradox

Yuval Noah Harari, "AI and the paradox of trust," YouTube Video, June 30, 2025 (06:11 and 01:11).
https://www.youtube.com/watch?v=8GaW36EfidI

The Currency of Progress

Sanjay Oberoi, "The Evolution Of Netflix: From DVD Rentals To Global Streaming Leader," Seat11A, December 3, 2024.
https://seat11a.com/blog-the-evolution-of-netflix-from-dvd-rentals-to-global-streaming-leader/

Oxford Executive, "Case Study: Netflix's Transition from DVD Rental to Streaming," November 24, 2024.
https://oxfordexecutive.co.uk/case-study-netflixs-transition-from-dvd-rental-to-streaming/

Terry Gross, NPR, "What Leaked Internal Documents Reveal About the Damage Facebook Has Caused," September 23, 2021.
https://www.npr.org/2021/09/23/1040081678/what-leaked-internal-documents-reveal-about-the-damage-facebook-has-caused

Atomic Mail, "Facebook Privacy Settlement: All You Need to Know," accessed November 28, 2025.
https://atomicmail.io/blog/facebook-privacy-settlement-all-you-need-to-know

Siobhan Moss, Social Media Curve, "47 Facebook Statistics, Facts & Trends For 2025," May 9, 2025.
https://socialmediacurve.com/facebook-statistics/

"The Top 25 AI Companies of 2025," The Software Report, April 23, 2025.
https://www.thesoftwarereport.com/the-top-25-ai-companies-of-2025/

References

Is AI Solving the Right Problem?

Michael Sandel, "Is AI Productivity Worth Our Humanity? with Prof. Michael Sandel," Your Undivided Attention Podcast, YouTube, June 26, 2025 (32:41).
https://youtu.be/L1U2haTj-Ws?si=J-twByI-YdYtGg1b

Emagine, "Key Trends Shaping 2025: AI agents lead the way," accessed November 28, 2025.
https://www.emagine.org/blogs/ai-agents-lead-the-way-key-trends-shaping-2025/

Vinciane Beauchene, Sylvain Duranton, Nipun Kalra, and David Martin, BCG, "AI at Work: Momentum Builds but Gaps Remain," June 26, 2025.
https://www.bcg.com/publications/2025/ai-at-work-momentum-builds-but-gaps-remain

Business Norway, "Smart Cities in Norway Enhance Quality of Life and Reduce Emissions," September 2, 2024.
https://businessnorway.com/articles/smart-cities-in-norway-enhance-quality-of-life-and-reduce-emissions

Molly Kinder, "Hollywood writers went on strike to protect their livelihoods from generative AI. Their remarkable victory matters for all workers," Brookings Institution, April 12, 2024.
https://www.brookings.edu/articles/hollywood-writers-went-on-strike-to-protect-their-livelihoods-from-generative-ai-their-remarkable-victory-matters-for-all-workers/

McKinsey & Company, "The pulse of nurses' perspectives on AI in healthcare delivery," October 1, 2024.
https://www.mckinsey.com/industries/healthcare/our-insights/the-pulse-of-nurses-perspectives-on-ai-in-healthcare-delivery

Baker McKenzie, "Navigating Labor's Response to AI," June 24, 2025.
https://www.bakermckenzie.com/en/insight/publications/2025/06/navigating-labors-response-to-ai

Between Chaos and Dystopia

Tristan Harris, "Is AI Apocalypse Inevitable? - Tristan Harris," After Skool, YouTube, June 17, 2025 (05:22).
https://www.youtube.com/watch?v=86k8N4YsA7c&t=2s

Chapter 2

Naming An Era

Seth Rosenberg, "Introducing Bots for Messenger," Facebook for Developers, April 12, 2016.
https://developers.facebook.com/blog/post/2016/04/12/bots-for-messenger/

Q. Vera Liao, Werner Geyer, Michael Muller, Yasaman Khazaeni, "Conversational Interfaces: A Review of Design and Research Efforts," (2020).
https://qveraliao.com/book2020.pdf

Mark P Taylor et al., "How organizations and consumers are embracing voice and chat assistants," Capgemini Research Institute, (2020).
https://www.the-digital-insurer.com/wp-content/uploads/2020/02/1573-Report_Conversational-Interfaces-1.pdf

Jason Yuan, "Reimagining Operating Systems Around Intent," MercuryOS, accessed October 13, 2025.
https://www.mercuryos.com/

Lennart Ziburski, “Intent-Driven Desktop Interface,” DesktopNeo, accessed October 13, 2025.
https://desktopneo.com/

“2025 Watch List: Artificial Intelligence in Health Care: Health Technologies [Internet],” Ottawa (ON): Canadian Agency for Drugs and Technologies in Health, March 2025.
https://www.ncbi.nlm.nih.gov/books/NBK613808/

Enrico Schaefer [reviewed by], “Recent Lawsuits Against AI Companies: Beyond Copyright,” Traverse Legal, May 19, 2025.
https://www.traverselegal.com/blog/ai-litigation-beyond-copyright/

Mobile Matters

Andrew Buck, “What Percentage of Internet Traffic is Mobile,” MobiLoud, July 3, 2025.
https://www.mobiloud.com/blog/what-percentage-of-internet-traffic-is-mobile

Chantal Line Carpentier [prepared by], “The Attention Economy,” United Nations Economist Network, United Nations, February 2025.
https://www.un.org/sites/un2.un.org/files/attention_economy_feb.pdf

Josh Howarth, “Smartphone Addiction Statistics,” Exploding Topics, April 13, 2025.
https://explodingtopics.com/blog/smartphone-addiction-stats

“Mobile Artificial Intelligence Market Size,” Nova One Advisor, May 2025.
https://www.novaoneadvisor.com/report/mobile-artificial-intelligence-market

Chris Arkenberg, Duncan Stewart, Gillian Crossan, Kevin Westcott, "On-device generative AI could make smartphones more exciting—if they can deliver on the promise," Deloitte Center for Technology Media & Telecommunications, November 9, 2024.
https://www.deloitte.com/us/en/insights/industry/technology/technology-media-and-telecom-predictions/2025/gen-ai-on-smartphones.html

"Apple Foundation Models: 2025 Updates," Apple Machine Learning Research, June 9, 2025.
https://machinelearning.apple.com/research/apple-foundation-models-2025-updates

Voice Search Statistics," Yaguara, May 21, 2025.
https://www.yaguara.co/voice-search-statistics/

Chapter 3

There's An App For That

"The Attention Economy," Center for Humane Technology, accessed October 13, 2025.
https://www.humanetech.com/youth/the-attention-economy

When Trust Was Honored

"History of Wikis," Wikipedia, accessed October 13, 2025.
https://en.wikipedia.org/wiki/History_of_wikis

"History of Wikipedia," Wikipedia, accessed October 13, 2025.
https://en.wikipedia.org/wiki/History_of_Wikipedia

Nathan Ingraham, "Apple Shutting Down iAd App Network at the End of June," Engadget, January 15, 2016.
https://www.engadget.com/2016-01-15-apple-iad-shutdown-june-30th.html

Andy Greenberg, "Apple's 'Differential Privacy' Is About Collecting Your Data---But Not Your Data," Wired, Jun 13, 2016.
https://www.wired.com/2016/06/apples-differential-privacy-collecting-data/

"A Timeline of Apple's Privacy Changes in Safari and iOS," Z2A Digital, August 1, 2025.
https://www.z2adigital.com/blog-content/a-timeline-of-apples-privacy-changes-in-safari-and-ios-infographic

Daniel Newman, "Apple, Meta and the Ten Billion Dollar Impact of Privacy Changes," Forbes, February 10, 2022.
https://www.forbes.com/sites/danielnewman/2022/02/10/apple-meta-and-the-ten-billion-dollar-impact-of-privacy-changes/

Ben Gilbert, "Facebook Blames Apple for $10 Billion Loss, Ad Privacy Warning," Business Insider, February 2, 2022.
https://www.businessinsider.com/facebook-blames-apple-10-billion-loss-ad-privacy-warning-2022-2

Mike Wuerthele, "'Privacy. That's iPhone' Ad Campaign Launches, Highlights Apple's Stance on User Protection," AppleInsider, March 14, 2019.
https://appleinsider.com/articles/19/03/14/privacy-thats-iphone-ad-campaign-launches-highlights-apples-stance-on-user-protection

Jesse Wiafe, "iOS 18 for App Marketers," Yodel Mobile, October 3, 2024.
https://yodelmobile.com/ios-18-for-app-marketers/

When Extraction Became The Model

Lauren Kaori Gurley, "Internal Documents Show Amazon's Dystopian System for Tracking Workers Every Minute of Their Shifts," Vice, June 2, 2022. **https://www.vice.com/en/article/internal-documents-show-amazons-dystopian-system-for-tracking-workers-every-minute-of-their-shifts/**

Bilal Kassem, "Senate Investigation Exposes Amazon Warehouse Injury Crisis," Pacific Workers, February 27, 2025. **https://www.pacificworkers.com/blog/2025/february/senate-investigation-exposes-amazon-warehouse-in/**

Emily Guendelsberger, "Amazon Treats Its Warehouse Workers Like Robots: Ex-Employee," Time, July 18, 2019. **https://time.com/5629233/amazon-warehouse-employee-treatment-robots/**

Jessica Mendoza and Ryan Knutson, "How TikTok Became the World's Favorite App," The Wall Street Journal Podcast (Transcript), February 20, 2023. **https://www.wsj.com/podcasts/the-journal/how-tiktok-became-the-worlds-favorite-app/0e3e20e8-9c39-4e34-beff-ca6cf058ddca**

Lookingbill V, Mohammadi E, Cai Y. "Assessment of Accuracy, User Engagement, and Themes of Eating Disorder Content in Social Media Short Videos," JAMA Netw Open. 2023;6(4):e238897. **https://jamanetwork.com/journals/jamanetworkopen/fullarticle/2803947**

Julianne Gabor, "The TikTok Algorithm Is Good, But Is It Too Good? Exploring the Responsibility of Artificial Intelligence Systems Reinforcing Harmful Ideas on Users," 32 Cath. U. J. L. & Tech 109 (2023). **https://scholarship.law.edu/jlt/vol32/iss1/6**

Scott Simon, "Parents in the Senate want new limits on social media to protect kids' mental health," Weekend Edition, NPR, May 6, 2023. **https://www.npr.org/2023/05/06/1174468842/parents-in-the-senate-want-new-limits-on-social-media-to-protect-kids-mental-hea**

The Sewell Setzer Case

"Garcia v. Character Technologies Complaint," October 23, 2024. **https://cdn.arstechnica.net/wp-content/uploads/2024/10/Garcia-v-Character-Technologies-Complaint-10-23-24.pdf**

Clare Duffy, "Teen Suicide and Character AI Lawsuit," CNN, October 30, 2024. **https://www.cnn.com/2024/10/30/tech/teen-suicide-character-ai-lawsuit**

Amanda Silberling, "Parents Sue OpenAI Over ChatGPT's Role in Son's Suicide," TechCrunch, August 26, 2025. **https://techcrunch.com/2025/08/26/parents-sue-openai-over-chatgpts-role-in-sons-suicide/**

Olivia Young, "Lawsuit: Character.AI Chatbot Colorado Suicide," CBS News Colorado, October 2, 2025. **https://www.cbsnews.com/colorado/news/lawsuit-characterai-chatbot-colorado-suicide/**

Rebecca Bellan, "Leaked Meta AI Rules Show Chatbots Were Allowed to Have Romantic Chats with Kids," TechCrunch, August 14, 2025. **https://techcrunch.com/2025/08/14/leaked-meta-ai-rules-show-chatbots-were-allowed-to-have-romantic-chats-with-kids/**

Rebecca Bellan, "California Becomes First State to Regulate AI Companion Chatbots," TechCrunch, October 13, 2025. **https://techcrunch.com/2025/10/13/california-becomes-first-state-to-regulate-ai-companion-chatbots/**

References

Chapter 4

A Model of Conscious Choice

Antti Saarnio, interviewed by Van Eiseman via video calls, October 15 and November 5, 2025.

Göte Nyman, Antti Saarnio, "New generation operating system to work with AI," Göte Nyman's (gotepoem) Blog, November 30, 2024. **https://gotepoem.wordpress.com/2024/11/30/new-generation-operating-system-to-work-with-ai/**

Venho.AI, "Human-Centric AI Operating System," accessed October 2025. **https://venho.ai/**

Jolla, "Mind2 AI Computer," accessed October 2025. **https://www.jollamind2.com/**

Chapter 5

When Presence Disappeared

Fandom Spotlight, "THE BREAKFAST CLUB 40 Year FULL Reunion – C2E2 2025," YouTube Video, June 30, 2025 (37:50). **https://www.youtube.com/watch?v=PEhWm4AKgR0**

Fabio Duarte, "Time Spent Using Smartphones (2025 Statistics)," Exploding Topics, June 5, 2025. **https://explodingtopics.com/blog/smartphone-usage-stats**

Wei Zhu, Ying Zhang, Yanzhi Lan, Xinqiang Song, “Smartphone dependence and its influence on physical and mental health,” Frontiers in Psychiatry, 16, 1281841, (2025).
https://doi.org/10.3389/fpsyt.2025.1281841

Sofia Ramirez, “Mobile Phone Usage Statistics 2025: What the Latest Data Reveals,” SQ Magazine, October 1, 2025.
https://sqmagazine.co.uk/mobile-phone-usage-statistics/

Human Connection Enhancement in Practice

“How AI Translation Can Increase Civic Engagement,” Worldly Blog, February 25, 2025.
https://www.wordly.ai/blog/civic-engagement

Dr. Rachna Jain, “AI and Empathy: Strengthening Client Connections Through Technology,” November 16, 2024.
https://www.linkedin.com/pulse/ai-empathy-strengthening-client-connections-through-technology-jain-h6qce/

“How Sephora’s Chatbot Transforms Beauty Shopping,” AgentiveAIQ Blog, August 18, 2025.
https://agentiveaiq.com/blog/how-sephoras-chatbot-transforms-beauty-shopping

Hao Wang, “Algorithmic Colonization of Love: The Ethical Challenges of Da ting App Algorithms in the Age of AI,” Techné: Research in Philosophy and T echnology (Online First), Department of Philosophy, University of Amsterdam, November 21, 2023.
https://philarchive.org/archive/WANACO-6

Jianwen Zheng, Justin Zuopeng Zhang, Muhammad Mustafa Kamal, Xiaoyang Liang, Ebtesam Abdullah Alzeiby, "Unpacking human-AI interaction: Exploring unintended consequences on employee Well-being in entrepreneurial firms through an in-depth analysis," Journal of Business Research, 196, 115406, (2025). **https://doi.org/10.1016/j.jbusres.2025.115406**

Frank Landymore, "Over 50 Percent of the Internet Is Now AI Slop, New Data Finds: Humans aren't finished... yet," Futurism, October 14, 2025. **https://futurism.com/artificial-intelligence/over-50-percent-internet-ai-slop**

Chapter 6

Trust-Centered Design in Practice

Sri Krishna, "Hugging Face Takes Step Toward Democratizing AI and ML," VentureBeat, September 27, 2022. **https://venturebeat.com/ai/hugging-face-steps-toward-democratizing-ai-and-ml-with-latest-offering%EF%BF%BC/**

Jacob Beningo, Beningo Embedded Group, "Running LLMs Locally for Embedded Development with Ollama," Design News, June 2, 2025. **https://www.designnews.com/artificial-intelligence/ollama-run-ai-models-locally-for-secure-cost-effective-embedded-development**

Paul Robichaux, "Practical AI: Make Your LLM Local with Jan," Practical365, September 9, 2025. **https://practical365.com/practical-ai-make-your-llm-local-with-jan/**

Paul Sawers, "Proton Launches 'Privacy-First' AI Writing Assistant for Email That Runs On-Device," TechCrunch, July 18, 2024. **https://techcrunch.com/2024/07/18/proton-launches-privacy-first-ai-writing-assistant-for-email-that-runs-on-device/**

Joe Foley, "So Adobe Firefly AI Isn't as Squeaky Clean as It Seemed," Creative Bloq, April 14, 2024.
https://www.creativebloq.com/news/adobe-firefly-trained-on-midjourney

Sara Merken, "Judge fines lawyers in Walmart lawsuit over fake, AI-generated cases," Reuters, February 25, 2025.
https://www.reuters.com/legal/government/judge-fines-lawyers-walmart-lawsuit-over-fake-ai-generated-cases-2025-02-25/

Mike Scarcella, "US judge approves 'novel' Clearview AI class action settlement," Reuters, March 21, 2025.
https://www.reuters.com/legal/litigation/us-judge-approves-novel-clearview-ai-class-action-settlement-2025-03-21/

Chapter 7

The Interface Never Intended

Denys A, "The 7±2 Rule: The Science Behind Our Cognitive Capacity," PsychoTricks, August 28, 2024.
https://psychotricks.com/7%C2%B12-rule-millers-law/

Charles Passy, "Before Google, you had to Ask Jeeves," MarketWatch, September 27, 2023.
https://www.marketwatch.com/story/before-google-you-had-to-ask-jeeves-6dba7837

"Conversational AI Market Size, Share | Statistics 2025-2032," Fortune Business Insights, December 15, 2025.
https://www.fortunebusinessinsights.com/conversational-ai-market-109850

Defining Seamless Integration

Matthias Stadler, Maria Bannert, Michael Sailer, "Cognitive ease at a cost: LLMs reduce mental effort but compromise depth in student scientific inquiry," Computers in Human Behavior, Volume 160, 108386, ISSN 0747-5632, 2024.
https://doi.org/10.1016/j.chb.2024.108386

Yang, Y., Zhang, Y., Sun, D. et al. "Navigating the landscape of AI literacy education: insights from a decade of research (2014–2024)," Humanit Soc Sci Commun 12, 374 (2025)
https://doi.org/10.1057/s41599-025-04583-8

Seamless Integration in Practice

MercuryOS, "Mercury OS is a speculative vision designed to question the paradigms governing human-computer interaction today," accessed November 2025.
https://www.mercuryos.com/

Desktop Neo, "Rethinking the desktop interface for productivity," accessed November 2025.
https://desktopneo.com/

Sonal Gupta, "5 Ways Equity Research Teams Use Hebbia to Drive Speed and Insight," Hebbia Blog, July 10, 2025.
https://www.hebbia.com/blog/5-ways-equity-research-teams-use-hebbia-to-drive-speed-and-insight

"Phenom Announces World's Most Advanced Applied AI to Solve Industry-Specific Hiring, Development and Retention Challenges, Eliminating the Gap Between Business Goals and HR Strategies," Phenom Newsroom, March 12, 2025.
https://www.phenom.com/press-release/phenom-announces-worlds-most-advanced-applied-ai-to-solve-industry-specific

Nicole Deslandes, "DPD disables 'sweary' AI chatbot," TechInformed, January 22, 2024.
https://techinformed.com/dpd-disables-sweary-ai-chatbot/

Nick Robins-Early, "McDonald's ends AI drive-thru trial as fast-food industry tests automation," The Guardian, June 17, 2024.
https://www.theguardian.com/business/article/2024/jun/17/mcdonalds-ends-ai-drive-thru

Aminu Abdullahi, "Taco Bell Scales Back AI Plans as Drive-Through Tech Faces Problems," eWeek, September 1, 2025.
https://www.eweek.com/news/taco-bell-ai-drive-thru-setbacks/

"NYC's AI chatbot criticised for advising businesses to break the law," Associated Press through Euronews, April 4, 2024.
https://www.euronews.com/next/2024/04/04/nycs-ai-chatbot-criticised-for-advising-businesses-to-break-the-law

Chapter 8

Anticipatory AI in Practice

Adams, R., Henry, K.E., Sridharan, A. et al, "Prospective, multi-site study of patient outcomes after implementation of the TREWS machine learning-based early warning system for sepsis," Nat Med 28, 1455–1460 (2022).
https://doi.org/10.1038/s41591-022-01894-0

Sida Peng, Eirini Kalliamvakou, Peter Cihon, Mert Demirer, "The Impact of AI on Developer Productivity: Evidence from GitHub Copilot," arXiv preprint arXiv:2302.06590, (2023).
https://arxiv.org/pdf/2302.06590

Lång K, Josefsson V, Larsson AM, Larsson S, Högberg C, Sartor H, Hofvind S, Andersson I, Rosso A, "Artificial intelligence-supported screen reading versus standard double reading in the Mammography Screening with Artificial Intelligence trial (MASAI): a clinical safety analysis of a randomised, controlled, non-inferiority, single-blinded, screening accuracy study," Lancet Oncol. 2023 Aug;24(8):936-944.
https://doi.org/10.1016/s1470-2045(23)00298-x

Heemsbergen, Luke, Radhika Gorur, Shiri Krebs, and Alexia Maddox, "Algorithmic Performance Management in Higher Education: Viva! 365 Ways of Surveillance," Surveillance & Society 22(2): 73-87. (2024).
https://ojs.library.queensu.ca/index.php/surveillance-and-society/article/view/15776/11579

Martin Kaste, "Facebook Increasingly Reliant on A.I. To Predict Suicide Risk," NPR - All Things Considered, November 17, 2018.
https://www.npr.org/2018/11/17/668408122/facebook-increasingly-reliant-on-a-i-to-predict-suicide-risk

Liv McMahon, "Meta to stop its AI chatbots from talking to teens about suicide," BBC, September 1, 2025.
https://www.bbc.com/news/articles/c2kzl79jv15o

Jason Winshell, "Drivers Protest Uber's 'Black Box' Fare System," San Francisco Public Press, July 10, 2025.
https://www.sfpublicpress.org/drivers-protest-ubers-black-box-fare-system/

Chapter 9

Intimate Gateways

Fabio Duarte, "Internet Traffic from Mobile Devices (July 2025)," Exploding Topics, July 10, 2025.
https://explodingtopics.com/blog/mobile-internet-traffic

"Mobile Fact Sheet" Pew Research Center, November 13, 2024.
https://www.pewresearch.org/internet/fact-sheet/mobile/

Alexandra Norris, Dorothe Singer, "Digital technology is unlocking financial inclusion," World Bank Blogs, July 17, 2025.
https://blogs.worldbank.org/en/developmenttalk/digital-technology-is-unlocking-financial-inclusion

Team Counterpoint, "GenAI Smartphone Shipments to Exceed 400 Million in 2025, Capturing One-third of Global Market," Counterpoint Research, March 19, 2025.
https://counterpointresearch.com/en/insights/genai-smartphone-shipments-to-exceed-400-million-in-2025-capturing-onethird-of-global-market

Mobile as AI Gateway in Practice

Vyom Ramani, "How to run offline AI LLM model on Android using MLC Chat," digit.in, July 14, 2025.
https://www.digit.in/features/general/how-to-run-offline-ai-llm-model-on-android-using-mlc-chat.html

Aiko - AI Speech-to-Text by Sindre Sorhus, download from the Apple App Store:
https://apps.apple.com/us/app/aiko/id1672085276

Whisper Notes Speech-to-Text by Yeonni Lee, download from the Apple App Store:
https://apps.apple.com/us/app/whisper-notes-speech-to-text/id6447090616

The AI Edge Authors, "On-Device Training with LiteRT," Google AI for Developers, (2024).
https://ai.google.dev/edge/litert/models/ondevice_training

Pat Lawlor, Jerry Chang, "Getting personal with on-device AI," Qualcomm ONQ Blog, October 11, 2023.
https://www.qualcomm.com/news/onq/2023/10/getting-personal-with-on-device-ai

"Improving Your Recommendations on Our Apps With AI at Meta," Meta, October 1, 2025.
https://about.fb.com/news/2025/10/improving-your-recommendations-apps-ai-meta/

Echo Wang, "Meta to use AI chats to personalize content and ads from December," Reuters, October 1, 2025.
https://www.reuters.com/business/media-telecom/meta-use-ai-chats-personalize-content-ads-december-2025-10-01/

Jon Keegan, "Life360 Sued for Selling Location Data," The Markup, June 1, 2023.
https://themarkup.org/privacy/2023/06/01/life360-sued-for-selling-location-data

Benjamin Preston, "Usage-Based Car Insurance Can Save You Money, but It Puts Your Data Privacy at Risk," Consumer Reports, August 21, 2025.
https://www.consumerreports.org/money/car-insurance/car-insurance-telematics-pros-and-cons-a5869096072/

David Nield, "How to Limit Galaxy AI to On-Device Processing—or Turn It Off Altogether," Wired, July 20, 2025.
https://www.wired.com/story/limit-galaxy-ai-to-on-device-processing-or-turn-it-off/

Chapter 10

The Price Being Paid

Dan Swinhoe, "Facebook to take up to 250MW in QTS' Hillsboro campus in Oregon," Data Centre Dynamics, September 24, 2021.
https://www.datacenterdynamics.com/en/news/facebook-to-take-up-to-250mw-in-qts-hillsboro-campus-in-oregon/

"One average megawatt is enough to power 796.36 Northwest homes for a year," according to The Northwest Power and Conservation Council.
https://www.nwcouncil.org/reports/columbia-river-history/megawatt/

Monica Samayoa, "Oregon Legislature passes 'POWER Act,' targeting industrial energy users like data centers," OPB, June 5, 2025.
https://www.opb.org/article/2025/06/05/oregon-data-centers-cryptocurrency-business-environment-power-electricity/

Defining Environmentally Responsible Innovation

Taiba Jafari, Olexandr Balyk, Lewis (Zhaoyu) Wu, James Glynn, "Projecting the Electricity Demand Growth of Generative AI Large Language Models in the US," Center on Global Energy Policy at Columbia | SIPA, July 17, 2024.
https://www.energypolicy.columbia.edu/projecting-the-electricity-demand-growth-of-generative-ai-large-language-models-in-the-us/

IEA (2025), "Energy and AI," IEA, Paris, Licence: CC BY 4.0.
https://www.iea.org/reports/energy-and-ai

Conscious Responsibility

Charlotte Trueman, “Microsoft has AI GPUs “sitting in inventory” because it lacks the power necessary to install them,” Data Centre Dynamics, November 03, 2025.
https://www.datacenterdynamics.com/en/news/microsoft-has-ai-gpus-sitting-in-inventory-because-it-lacks-the-power-necessary-to-install-them/

“Fingrid secures transmission capacity for growth in electricity consumption and new industrial investments – restrictions on new energy storage facility connections continued in southern Finland,” Fingrid Press Release, September 26, 2025.
https://www.fingrid.fi/en/news/news/2025/fingrid-secures-transmission-capacity-for-growth-in-electricity-consumption-and-new-industrial-investments--restrictions-on-new-energy-storage-facility-connections-continued-in-southern-finland/

Responsible Innovation in Practice

Kevin Brehm, Mary Tobin, “Virtual Power Plant Flipbook,” VP3 in conjunction with RMI, (2023).
https://rmi.org/wp-content/uploads/dlm_uploads/2024/06/VP3_flipbook_v1.1.pdf

Sasha Luccioni, Régis Pierrard, “Reduce, Reuse, Recycle: Why Open Source is a Win for Sustainability,” HuggingFace Community, May 7, 2025.
https://huggingface.co/blog/sasha/reduce-reuse-recycle

“Syntiant Brings AI Development to Everyone, Everywhere with Introduction of TinyML Platform,” Syntiant Corp Press Release, September 29, 2021.
https://www.syntiant.com/news/syntiant-brings-ai-development-to-everyone-everywhere-with-introduction-of-tinyml-platform

"Syntiant Core 2 Achieves Lowest Power Results in MLPerf Tiny v1.2 Benchmark Suite," Syntiant Corp Press Release, April 22, 2024. **https://www.syntiant.com/news/syntiant-core-2-achieves-lowest-power-results-in-mlperf-tiny-v12-benchmark-suite**

Malina McLennan, "The digital drill: How big oil is using AI to speed up fossil fuel extraction," Global Witness, September 21, 2023. **https://globalwitness.org/en/campaigns/fossil-fuels/the-digital-drill-how-big-oil-is-using-ai-to-speed-up-fossil-fuel-extraction/**

Cameron Wade, et al., "Electricity Grid Impacts of Rising Demand from Data Centers and Cryptocurrency Mining Operations," The Open Energy Outlook (OEO) Initiative, June 2025. **https://energy.cmu.edu/_files/documents/electricity-grid-impacts-of-rising-demand-from-data-centers-and-cryptocurrency-mining-operations.pdf**

Miguel Yañez-Barnuevo, "Data Centers and Water Consumption," Environmental and Energy Study Institute (EESI), June 25, 2025. **https://www.eesi.org/articles/view/data-centers-and-water-consumption**

Shaolei Ren, Amy Luers, "The Real Story on AI's Water Use–and How to Tackle It," IEEE Spectrum, September 10, 2025. **https://spectrum.ieee.org/ai-water-usage**

The Environmental Imperative

Joe Schulz, "Data center 'statewide guardrails' proposed under Wisconsin bill," Wisconsin Public Radio WPR, November 10, 2025. **https://www.wpr.org/news/data-center-wisconsin-guardrails-proposed-bill**

References

"$64 billion of data center projects have been blocked or delayed amid local opposition," Data Center Watch Report, Accessed November 20, 2025. **https://www.datacenterwatch.org/report**

IEA (2023), "Net Zero Roadmap: A Global Pathway to Keep the 1.5 °C Goal in Reach," IEA, Paris, Licence: CC BY 4.0 **https://www.iea.org/reports/net-zero-roadmap-a-global-pathway-to-keep-the-15-c-goal-in-reach**

www.ingramcontent.com/pod-product-compliance
Lightning Source LLC
Chambersburg PA
CBHW030509210326
41597CB00013B/842

* 9 7 9 8 9 9 3 0 2 3 3 0 4 *